P9-CKF-825

The Humane Economy

ALSO BY WAYNE PACELLE

The Bond

The Humane Economy

How Innovators and Enlightened Consumers Are Transforming the Lives of Animals

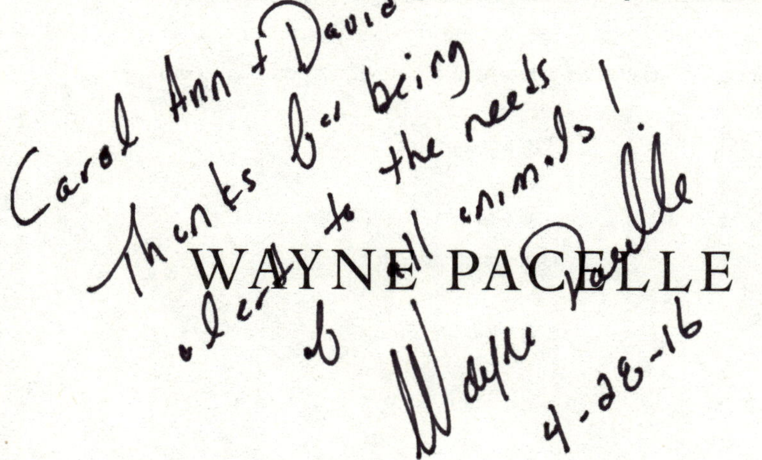

WAYNE PACELLE

WILLIAM MORROW
An Imprint of HarperCollins*Publishers*

For information address HarperCollins Publishers,
195 Broadway, New York, NY 10007.

HarperCollins books may be purchased for educational, business, or
sales promotional use. For information please e-mail the
Special Markets Department at SPsales@harpercollins.com.

FIRST EDITION

Library of Congress Cataloging-in-Publication Data
has been applied for.

ISBN 978-0-06-238964-0

16 17 18 19 20 NMSG/RRD 10 9 8 7 6 5 4 3 2 1

Dedicated to my friend Audrey Steele Burnand,
a one-of-a-kind champion of all animals

Contents

Introduction

"And hence it is, that to feel much for others and little for ourselves; to restrain our selfishness and exercise our benevolent affections, constitute the perfection of human nature."

—ADAM SMITH, *The Theory of Moral Sentiments*

THIS IS A REMARKABLE era in history. Never has there been such widespread and determined concern for the welfare of animals. And yet, we humans mete out pain and toil on animals on a scale unseen, indeed unimagined, before.

What's the takeaway from this contradiction? One word: hope.

Exploiting animals is a practice under siege—whether in puppy mills and pet shops, circuses and marine parks, factory farms and slaughterhouses, mink farms and fur salons, and primate laboratories and cosmetic testing facilities. Concern for animals is ascendant. And today there's a fast growing, often surprising, hugely promising, and largely unstoppable force for animal welfare, and it's revealing itself in a thousand varying forms. Welcome to the humane economy.

If you are part of the old, inhumane economic order, get a new business plan or get out of the way. You're already in danger of being too late. Every day there is less room in our civic conversations for discredited ideas about animals existing for whatever use we humans concoct, and less tolerance for self-serving rationalizations for calculated cruelty. Those old ways of thinking are being squeezed into oblivion from two sides.

On one hand, there's a groundswell among consumers who not only believe that animals matter but also put those principles into action and make choices that drive change in the marketplace. This freshly turned economic soil nurtures legions of hungry entrepreneurs who are imagining better ways to produce goods and services that do less or no harm to animals. These visionary entrepreneurs are enlisting scientists, economists, engineers, designers, architects, and marketers to the cause of providing food, clothing, shelter, healthcare, research techniques, and even entertainment, without leaving a trail of animal victims behind. This economic revolution is nothing short of astonishing in depth, breadth, and potential.

On the other hand, the humane economy is being propelled just as surely by people who are not intentionally out to end suffering but whose innovative work moves us in that direction anyway. It was primarily Henry Ford and not American Society for the Prevention of Cruelty to Animals (ASPCA) founder Henry Bergh who was at the wheel in dramatically reducing cruelty to horses in the nineteenth and early twentieth centuries. Ford's invention of the mass-produced automobile was not motivated by any special desire to spare the beasts of burden. But that was one lasting outcome. And it happened in a mere eyeblink of history. Few who lived in a nineteenth century American city would have thought it possible for such a rapid conversion from animal to machine transportation to occur. In fact, our language is still hitched to animal transport and hasn't even caught up to that distant revolution. To this day we measure our cars' engines by horsepower.

As recently as the early twentieth century, we tied messages to pigeons and sent them off into the sky for delivery. Before that, the Pony Express had a brief run in the nineteenth century. Today, Federal Express and DHL can deliver packages almost anywhere overnight with payload capacity and navigation systems that any pony or pigeon would envy. Amazon is experimenting with delivering books and other products by drone. And of course, with just a few keystrokes, we can download books to an electronic reader or send electronic messages and documents of any size in seconds across the planet.

Today, with the carrier pigeon and, to a considerable degree, the working horse in our rearview mirror, we must wonder what other animals might be spared their particular burdens by the powerful forces of innovation. Given the intensity and scale of animal exploitation today, in so many different sectors of the economy, why wouldn't we make urgent efforts to harness innovation to make cruel uses of animals obsolete? Our human creativity and our increasingly alert moral temperament make this a world rife with opportunity, one that's swirling with the spirit of reinvention and social, technological, and economic reform.

In *Capitalism, Socialism and Democracy*, the eminent economist Joseph Schumpeter described capitalism as "a perennial gale of creative destruction," the process by which entrepreneurs and innovators introduce new goals, new means of production, and new products in support of their visions. The old businesses often make apocalyptic predictions about the new approaches. But changes in business attitudes and practices, as Schumpeter noted, drive growth and are the lifeblood of the economy: businesses that do not adapt are left behind, while innovators claim a larger share of the market.

When it comes to the humane economy, making money *and* doing good is precisely the point. If ideas about compassion are going to prevail, they must triumph in the marketplace. We can produce high-quality goods, services, or creative content and also honor animal-protection values in the process. We can feed the world's surging population without resorting to extreme confinement of animals. We can validate the safety of cosmetics and chemicals without poisoning mice or rabbits. We can solve human–wildlife conflicts without resorting to bursts of violence.

Just about every enterprise built on harming animals today is ripe for disruption. Where there is a form of commercial exploitation, there is an economic opportunity waiting for a business doing less harm or no harm at all. Factory farming, for example, is the creation of human resourcefulness detached from conscience. What innovations in agriculture might come about by human resourcefulness guided by conscience?

With this book, I ask you to join me in meeting some of the pathfinders in the twenty-first century's humane economy, the people who are helping to usher in a series of transformations that will rival changes we've seen in the transportation sector within the last century or in information technology within the last two decades. Some of the biggest names in egg and pork production—once synonymous with intensive confinement of animals and part of the old, inhumane economic order—are tearing out the cages and crates. They're now converts and contributors to the humane economy. I'll show you how visionary entrepreneurs are at the leading edge of a tectonic shift in food production and retail—as twenty-first century business leaders and their customers demand that industry do better.

For those who want to take the animals out of the equation entirely, we'll go behind the scenes with the people cracking the code. They're creating facsimiles of eggs and chicken, with the taste and texture of the real thing but none of the cruelty. In a blind taste test, you'd be hard pressed to distinguish them, but when it comes to a moral test, there's no comparison.

Two penniless street performers had a vision of entertaining people by showcasing beautifully choreographed feats of human strength and agility; and their company, Cirque du Soleil, has now made competitors featuring dancing elephants or snarling tigers perfectly outdated and archaic. While the Cirque du Soleil founders didn't explicitly have animal welfare on their minds when developing their new enterprise, Betsy Saul was all about saving lives when she developed Petfinder.com. Her virtual shelter has helped millions of people in the market for a dog or cat find the pet of their dreams and save lives in the process.

And it's not just the entrepreneurs. Scientists are part of this new humane economy, too, including several doing their best to perfect growing meat in a lab, without raising a full-bodied creature with a heart and brain. I'll take you out on the open ranges of Colorado's Sand Wash Basin, where the fertility control work pioneered by Jay Kirkpatrick offers the prospect of saving American's wild horses and providing a solution to satisfy key stakeholders who've never seen eye-to-eye on the management decisions. And I'll tell you about reformers from within science, such as National Institutes of Health Director Francis Collins, who played the central role in ending the era of using chimpanzees in invasive experiments and is now calling into question the reliability of animal tests for millions of mice, rats, and rabbits and urging his fellow scientists to embrace alternative methods where they can.

While we celebrate the innovators and the scientists, you'll also meet the investors—the people who recognize that capital drives the humane economy, producing profits for society alongside a range of other social benefits. You're unlikely to see headlines about billionaire Jon Stryker, but he's putting millions into protecting our closest living relatives in nature—chimps and other great apes—while Microsoft co-founder Paul Allen is financing anti–wildlife trafficking campaigns in order to save endangered species. Both men realize that elephants, gorillas, and other African wildlife are worth more alive than dead, and their investments return profits to people who need the income most and provide local people an incentive to join in saving them too.

The adopters and the emulators also are crucial to the humane economy. The smartest of businesses mimic and even improve upon the work of innovators who have shaken up their field and upended conventional thinking. When Whole Foods Market adopted a new look and feel to its stores and started offering organic foods and humanely sourced animal products, it didn't take long for competitors to start changing their offerings and their aisles. When one fast-food restaurant goes cage free or crate free, others in the sector want to get in on the act too. When there's a big new idea, there's first a recoil and maybe a reverberation followed by an adjustment or a correction or two; and then, if it works, broad acceptance—and later, we wonder how we ever managed to do things the old way.

The humane economy is not some abstraction or far-off concept, partly because animals are all around us. So many of the changes afoot will touch your life and that of the people you know. Indeed, you are—or will be—driving many of these changes, whether it involves the food you eat, the pets you keep, the household products you buy, or the films or wildlife you watch. If we seize the opportunities now available to us—whether as first adopters or those who join the parade of progress—we can help shape the market and accelerate transformational changes for animals throughout the global economy.

Economic theory assumes that people act rationally, according to perfect information, but where animal-use industries are concerned there has long been a world of difference between theory and reality. So often we don't really know how an animal product was made, and sometimes we don't want to know—a stance that can be described in various ways, but cannot be called rational. Even to call the attitude "self-interested" misses a larger point, since it cannot be in anyone's interest to act in ignorance, or to make choices that might well go against our conscience if we knew more.

In the Information Age, awareness is spreading and with it crucial knowledge that cannot be unlearned about the suffering endured by animals, once largely unquestioned, in human enterprises. Reality is becoming harder to hide—which is one reason why factory farmers, for example, are so desperate to outlaw the mere taking of unauthorized photographs of the things they are doing every day. When a company's greatest fear is a knowledgeable, ethically alert customer, that company has problems that won't go away. Any economist will tell you that when new, relevant information is acquired about the supply side, then people will adjust their expectations on the demand side. This is happening throughout our economy, as more of us ask questions and act on the answers. And one by one, cruel industries find themselves on the wrong side of a market that is changing, fundamentally, forever, and for the better.

We can, on a personal and societal level, shed animal cruelty, displacing the animal exploitation economy of yesteryear and brushing aside the dusty arguments and political machinations of those who cling to it. The entrepreneurs and business leaders and scientists you'll read about here are working on solutions. They are getting an assist from the many groups and individuals agitating for change and calling upon lawmakers, judges, prosecutors, and corporate leaders to embrace a new humane standard.

One thing is for sure: we need not accept the idea of routine cruelty in agriculture, entertainment, wildlife management, or any other part of our economy and culture. Together, by adopting new standards through political channels and reinforcing what business leaders are doing and ready to do, we can create a new normal when it comes to our human relationship with animals. Here, in this humane economy, human ingenuity meets human virtue, and we discover at last that we can have it both ways—a better world for us and for animals too.

Come see.

CHAPTER ONE

Pets and GDP (the Gross Domesticated Product)

"When we have grasped the great central fact about animals, that they are in the full sense our fellow-beings, all else will follow for them."

—HENRY SALT, *The Story of My Cousins*

The Beagle Has Landed

Like all mutts, my beagle mix Lily is an original. Though beagles are famously ravenous, with no real off switch to their appetite, Lily's enthusiasm for food has been checked by an inexplicable fear of her dog bowl. To overcome this phobia, we've tried bowls of every shape and size—from round, square, and shallow to paper, ceramic, and even fine china. We've been to boutique pet stores and searched the shelves for just the right platter or plate. But no matter the dish, she approaches it warily, craning her neck to reach the food, her weight pressed almost entirely on her back legs, her eyes darting nervously. She's like an antelope at a watering hole, fearing crocodiles lurking below the surface and lions hidden in the reeds—except that she's on the fifth floor of a condo in Washington, D.C., and the only other animal around is a ten-pound tabby named Zoe, who at her worst is known to give Lily a playful swat every now and again. By trial and error, we've learned that our dog will eat her food only if it's placed on, of all things, a paper towel. It's no matter to Lily that the towel is supposed to go on the mess, not the other way around.

She's better at walks, but even then, Lily's not without her eccentricities. In that way of beagles, she can be stubborn, refusing to move in any direction not to her liking. Then there's the matter of her—how to put it?—"earodynamics." She controls her big, floppy ears like a pilot working the wings of a plane, and when she's got a prance in her step and she's facing the breeze, the flaps go up. Smiles wash over the faces of people who walk past us. A few have likened our Lily to a canine version of the Flying Nun.

Often, too, she'll halt, stiffen, and start to move slowly and mechanically, as if in some full-body cast. Then with all the litheness of Frankenstein's monster, she walks in a semicircle around some unseen obstacle. It looks like she's delicately stepping around an ancient burial ground, too sacred to be trod upon by even a single paw. After that little ritual, she loosens up again, resumes her prance, and the ears go out again. All is good.

In the parlance of dog psychology, she's exhibiting submissive behavior when she comes across a urine mark from a dog: "This is some other dog's territory, and I don't want to antagonize him, so I'll stay clear of it." That would be the general idea if I could read her canine mind. Of course, the other dogs wouldn't much care if Lily glided right past their signature traces, and there's not much use to marking in the first place. More than thirty thousand years since their domestication, that part of their wild nature is still at work, and there's no explaining to them that nobody's really controlling territories.

It's been two years since my wife Lisa and I had the great fortune to adopt our sweet little girl, after sad and mysterious circumstances placed Lily in a government-run shelter in Spotsylvania County in the Virginia exurbs outside of Washington. After a few days in the shelter, with no owner coming forward and no interest from adopters, she was placed on a euthanasia list. Not enough space. Soon, she'd receive an injection of sodium pentobarbital and go lifeless before being placed in a trash bag and dumped in an incinerator. Ashes to ashes, dust to dust, without pause for a burial ceremony, nor anyone noticing that a little life had been taken before her time.

Instead, Lily made her way into our lives and our hearts, and she has brightened them every day since. The path she followed also makes her a living, breathing, ear-flapping, tiptoeing example of the second chance that can come to every dog in need, in a new and more hopeful era when we can finally put the euthanasia rooms and incinerators behind us. Gradually, all of that ugliness is giving way to a more humane, transparent, and rational approach—and so many other second chances. For nearly three million healthy dogs and cats still euthanized each year, that day can't come soon enough.

Lily's journey into our lives began with a band of volunteers with a group called Lost Dog and Cat Rescue. They keep watch on lists of shelter dogs and cats about to be euthanized throughout the Washington, D.C., area and in nearby states, and pull out a good number of them (rescues often take animals from shelters and promote adoptions through their own channels). Lost Dog volunteers saved Lily and transported her to their no-kill facility in northern Virginia.

After Lily got settled, the group took her to a veterinary clinic, where she got a checkup and medicine for problems with her stomach and ears and also her case of Lyme disease. Then, as a matter of organizational policy, a veterinarian spayed her. Volunteers took her on walks at adoption events and tried to socialize her, since she seemed so timid and in need of confidence building. After all of that care, they put her picture up on their website, announcing she was ready and waiting for adoption.

Lost Dog is under pressure—life-or-death pressure—to adopt out dogs and cats or foster them, so they can create more space to pull more animals from the euthanasia lists. That's how Lily got on the adoption circuit. But despite her adorable looks and a manageable thirty-pound frame—suited for city, suburban, or rural living—she was passed over at nine successive adoption events. "Good grief," the Charlie Brown in me has said so many times about the people who overlooked our little beagle: a wonderful girl, with a sweet disposition, those cute floppy ears, and still no takers. It makes you wonder how many other perfectly adoptable animals meet their demise every day, passed over because they couldn't score a perfect ten in that first encounter.

Her age didn't help (she was five or so, according to a vet who examined her teeth), since most people want a puppy and not an adult dog. It's not hard to see the appeal of an adorable puppy clambering all over you, and it's understandable to enjoy those precious early months of life. But Lisa and I were looking for an older dog precisely because they get passed over so routinely. Many of these dogs had known love and affection at some point in their lives, only to be relinquished when circumstances suddenly changed for the people they depended on. Maybe their caretakers had moved to an apartment with a no-pets policy, or perhaps they lacked the attention and patience to work through some irritating or destructive canine behavior. Lisa and I remind anyone who will listen that these older dogs, despite their dings and scrapes and even an occasional snaggletooth, seem especially grateful for the second chance. Some years ago, my friend Matthew Scully, an author and former presidential speechwriter, adopted an Australian shepherd mix named Herbie, who at thirteen years had plenty of gray across his muzzle. Weighing about seventy pounds and having arthritis, he was clearly in his twilight years. As it turned out, Matthew and his wife Emmanuelle had two years and a few months with Herbie before the old boy fell ill and passed. It was Herbie's "glorious prime," as Matthew puts it, filled with the special affection and gratitude of a senior dog. And it's an experience that both Matthew and Emmanuelle still cherish.

Some dogs just don't do well during their adoption auditions. When she met us, Lily acted skittish and distant, perhaps because she was unsettled by all the people and dogs pawing at her. She's not unlike a lot of other rescue dogs thrust into an unfamiliar environment, and it asks a lot of them to constantly exhibit best-in-show behavior. That's one reason why animal advocate Aimee Sadler developed a program calling Playing for Life, where shelter dogs are grouped in a play-yard for social time. As homeless dogs arrive and adoptees leave, they race, tangle, chase, work out pent-up energy, and figure out something of an ever-shifting social order. Playing for Life is designed to keep dogs from going "cage crazy"—a widespread issue in shelters where dogs, confined for long periods, bark and jump with abandon when they see someone. Too much eagerness, as with any first date, tends to be a turnoff. Dogs in these programs are calmer and better behaved, less hyperactive and high-strung—improving their prospects of going home with someone.

Lily, however, was anything but high-strung. She was meek as could be, with ears pulled back, eyes wide, and tail tucked tightly to her hindquarters. And when we first met her, despite the efforts of volunteers to prep her, she didn't exactly nail the audition. Moments after our eyes met, the poor girl had a diarrhea accident right at our feet. Talk about first impressions. This sort of an intestinal event would cause a lot of folks to hold their nose and avert their gaze. After that reflex, they'd likely get to thinking about their floors and carpets, the dog's training needs, and the less appealing but essential responsibilities of pet care. The Lost Dog folks said that Lily's diarrhea was brought on by stress and maybe some sort of stomach bug or inflammation. They played it cool with Lisa and me, but were doubtless thinking that Lily was about to go zero for ten.

After a quick cleanup, we calmed her down and coaxed her into playing a little. We got her on her back for a belly rub, and noticed right away that Lily had protruding nipples. She'd been a mama, with the puppies long gone. Perhaps she'd been a breeder in a puppy mill—entirely possible, since Virginia is one of the East's major mill states. She also had a couple of broken teeth, a second detail supporting that theory because severely confined and frustrated mill dogs often gnaw on the wires of their cages. They want out, and self-destructive behavior is their response to despair and isolation. Mill operators would never pay for any dental work—not even pulling broken teeth.

Lily also had scars on her legs, so if she wasn't a puppy-mill dog maybe she was a hunting hound who'd had some run-ins with barbed wire fences in pursuit of rabbits or other small prey. In fact, whenever I take her on a walk in a suburban or rural place and she gets locked on to the smell of a rabbit, she reverts back in a flash to her scent-hound mode—nose low to the ground, quickstepping left and right, and charging ahead, with that incredible sniffer sending lightning-fast instructions to her brain and legs. Not much can sidetrack her when she's in the zone, including my yanking on her harness. It's not hard to picture Lily, in her prior life, barreling over or scampering under a fence and picking up gashes on her legs. But while she has the nose and frame for hunting, she was almost certainly disqualified for other reasons—mainly because Lily hates loud, sharp noises. If any dog would run in exactly the wrong direction after hearing the crack of a shotgun or rifle, it's Lily. (We learned two months after we got her, when July Fourth came around, to drown out the sound of fireworks by turning up the television and running the dishwasher and the washer and dryer.) Rural Virginia shelters are full of discarded hounds, because it's a state where chasing animals with packs of dogs is still a tradition—and there's no law or code of conduct to prevent hunters from treating dogs as expendable if they don't perform. We had one more tidbit for the "discarded hunting dog" theory: Lily had untreated Lyme disease, which is more common to animals living outside, as hunting dogs typically do.

Adding to Lily's litany of health problems were eye and ear infections. That was just neglect and inattention, and her prior owners had undoubtedly been guilty of both. There wasn't anything about her condition that time, patience, and some medical attention couldn't soothe or remedy. But we'd never be sure what Lily had endured in her past, despite our best efforts to reconstruct the backstory and to undo the trauma. Our only certainty was our desire to fill the next chapter of this dog's life with contentment and love. Her age, health woes, and timid behavior didn't disqualify her for us. Just the opposite: we wanted to pull her close and see that nothing bad would befall her again. Happily for us and for Lily, the Lost Dog volunteers had found her just in time.

As it turned out, Lily would soon be joined in our household by another stray who sometimes puts her good nature to the test. Our cat Zoe traveled a straighter path to our door in the way many homeless cats do. Most people go out and find a dog when they want to bring one into their lives. Cats, on the other hand, typically find their new keepers. Zoe—who came to us with uncertain age but undeniable beauty—found me when I was out with Lily on an early morning jaunt.

When I first saw her, she appeared to be walking behind a pedestrian. A small number of people actually walk their cats, and I figured this was one of those unusual cases—a guy with his feline, off-leash for a stroll. But when I circled back after my loop with Lily, a young woman out for a jog was trying to coax the kitty to come to her. The man I saw earlier had disappeared, and it was clear that this cat, at least for the moment, was on her own. It didn't take much persuasion, and soon the cat was in the jogger's arms.

I made a lifeline call to Lisa for a cat carrier. Minutes later, she came down in her pajamas with sneakers that didn't match and carrier in hand. We quickly filled it with the wide-eyed stray. The nice jogger said she would consider taking the cat, but I knew she'd face a formidable rival in Lisa. The jogger took the cat to the local veterinarian, and for the moment, at least, the three of us tabled the custody question.

The veterinarian discovered the stray had an unregistered microchip. The clinic posted flyers. We posted a picture on several websites and called the Washington Humane Society lost pet hotline. We checked all the outlets daily, but nobody came forward to claim her.

With the thought that possession is 90 percent of the law, Lisa picked up the cat from the clinic and brought her home. I had the task of calling the very kind-hearted jogger, who, I learned, is a government lawyer. She grumbled a bit, but was too polite to cross-examine me, graciously agreeing that we could keep "Zoe," as we named her. The next time our paths crossed, I mentioned that Zoe is wont to pounce on me at night, and even tries to sleep on our necks sometimes. I told her that while she may have lost the cat, she's gained hundreds of hours of peaceful sleep.

Zoe's a character. She tears around the apartment without provocation. When I come home, she greets me at the door each evening and tries to sneak out to explore the hallway. I let her have her little taste of freedom, before I grab her and bring her back inside.

In the apartment, she's got plenty to keep her busy. Zoe has a big box full of toys, and when we take them out, she will play without stopping—she's the only cat I've known who pants because she's winded from all the stalking, pouncing, and racing around.

Like Lily, Zoe's got her share of quirks too. She chews on metal, for instance. The spout of our watering can looks like it's been through a meat grinder. But maybe she knows something we don't about dental care. When we take her to the veterinarian, he always remarks on what a beautiful, white, sharp set of teeth she has.

Smart Ideas at PetSmart and Petco

If you had asked me years ago where to get a dog, a pet store would not have been on my list. They were, in my long-held view, spruced-up, sanitized distribution outlets for cruel puppy mills, and a huge part of the problem for dogs in this country. At the mills, the puppies are separated from their mothers at a young age and forced to fend for themselves, often in extremes of heat and cold, before they're shipped off for sale at seven or eight weeks. The mothers have it worse; they're impregnated every heat cycle and often confined to cramped, filthy cages that they never leave. With no one to walk them, and no play-yard, they never touch grass or feel sunshine, and rarely get any exercise. They never feel loving touches from a human hand, and veterinary care is out of the question. If illness claims the life of a breeding mom, the puppy mill operators will just breed a new one and refill the cage.

The Humane Society of the United States (HSUS) and other animal-protection groups are making progress against these mills by supporting reforms in law, pressuring pet stores to stop selling mill puppies, and advocating pet adoption. Even so, thousands of these ramshackle operations still churn out dogs in various corners of rural America with the dogs confined in kennels and hidden from sight far from the road, nestled behind some barn or shed. Sometimes the dogs are actually crammed inside the sheds, hidden away and denied even sunlight, fresh air, or the least scrap of human kindness. As with so many other forms of mistreatment of animals, the abuse occurs in a cordoned-off, faraway place. Because people are so disassociated from it, they typically don't connect the filthy breeding dens in the mills and the spiffed-up sales cubbies in the pet stores—but they're all part of the same enterprise. It's one long supply chain, even as the mill owner, dog broker, and retail seller divide up the profits from all of this pain.

PetSmart and Petco were once the end sellers, primary conduits for the puppy mill industry. But in the mid-1990s, both companies, growing rapidly by providing all manner of supplies to pet owners, made a strategic decision: They said "no" to puppy mills and shed the entire practice of selling dogs and cats. Instead, they threw open their doors to local rescue groups and shelters, giving them access to the foot traffic in their stores. PetSmart and Petco don't make a dime on the adoption transactions—the fees go directly to the shelters and rescue groups. It was a new economic model for pet stores, centered on doing right for the animals and leveraging the emotional power of the adoption experience to earn and keep the loyalty of customers. Where better to get supplies for the pet you love than the place that brought that animal into your life?

For both companies, the decision to make a clean break from the mills was good for the dogs and cats and also great for business. Partnering with shelters and rescues brought more people to their stores and built lots of community goodwill. On the day we met Lily, Lisa and I set the navigation system for PetSmart with the sole purpose of finding a new companion. Even if we hadn't come home with a dog, we'd have left the store feeling very grateful for its adoption programs. And whether you take home a cat or a dog—or now, even a rabbit—at Petco or PetSmart, you can't turn in any direction without seeing supplies everywhere: kennels, leashes and collars; beds, sweaters, heating pads, and boots; food and water dishes, pet food, and treats; wee-wee pads and flushable cat litter; doggie seat belts, life preservers, and doggles (goggles for dogs); scratch pads; and toys—lots of toys. PetSmart reports that people who adopt from the store—euphoric in claiming their new family member and perhaps eager to wipe away any past troubles or discomforts the animal may have known—spend five times as much at the store as regular customers. That's some multiplier effect. And it doesn't end with a single shopping spree. Lisa and I love returning to that PetSmart store because it reminds us where Lily started her new life, and first brought such happiness into ours.

PetSmart's and Petco's adoption programs have been successful beyond anyone's expectations. They are each adding dozens of stores a year, and between them now have more than 2,500 outlets. Their main competition, apart from each other, isn't coming from the small pet stores any longer, but from the big box stores—mainly Walmart, Costco, and Target, which themselves have added major pet supply sections. Banfield Pet Hospitals has co-located 800 veterinary clinics—employing 5,000 veterinarians—right inside the PetSmart stores. Total sales within the US pet sector—meeting the needs of 170 million dogs and cats, along with more than 150 million other pets in our homes—are about $60 billion. In December 2014, as a sign of the times, BC Partners, a London-based firm, acquired PetSmart for $8.7 billion in the largest global private equity deal of the year. That's right: the biggest private equity acquisition in all of 2014 was not a company involved in finance, energy, pharmaceuticals, or mining, but pet products.

The in-store adoption program for the two pet supply giants has been transformative for the humane movement. With their shelter and rescue partners providing the animals, PetSmart and Petco have helped to transfer more than eleven million dogs and cats into loving homes—a game-changing number. They add more than half a million to that total every year. Many stores now have cat adoption centers that allow the kitties to live there round the clock, essentially serving as full-service shelters for these animals. And imagine if those 2,500 stores had been selling puppy mill dogs for these last two decades? Not only would the companies not have contributed millions of dogs to the adoption pipeline; rather they would have sold millions of puppies from mills instead, creating havoc for shelters and profits for mills. By doing something good in place of something bad, they've attacked the problem on a scale that is transformative for the cause.

Looking back, it may seem like an obvious marketing boon for these companies to pursue dog and cat adoptions exclusively. But the decisions of PetSmart and Petco were revolutionary at the time and disrupted the conventional pet store model. It required them to forgo live sales of dogs and cats—the animals at the very center of their enterprise and a prime lure for customers. The live sales were no loss leader, even as a once-in-a-lifetime transaction, because the stores made a hefty profit on each dog sold—at prices often reaching $1,000 or more.

By cutting ties with the puppy mill industry, and offering help to animals in need of homes, PetSmart and Petco became problem solvers, not problem makers, in relation to animal homelessness. Pet lovers are right to feel good about these companies, and all that customer loyalty is well deserved.

Right around the same time that PetSmart and Petco switched their model—providing thousands of outlets for adoption of homeless animals—humane innovators developed another business idea. It would prove every bit as game-changing in the world of animal sheltering. To put it in basic terms of supply and demand, shelters and rescues had an oversupply of goods (homeless dogs and cats) but not the right mix of goods for would-be pet owners (dogs of different sizes and breeds, and cats of different colors). Some urban shelters were overloaded with pit-bull-type dogs, while southwestern shelters had too many Chihuahuas, and southern shelters had lots of hounds. The inventory (dogs of every shape and size) was there, just not in the right places.

That's where two very forward-looking people came into the picture. On New Year's Eve, 1995, Betsy Banks Saul, then an urban forester, and Jared Saul, a medical resident, were on their way to dinner, talking about the challenges of pet adoption. They'd also been discussing a cool new thing called the World Wide Web. What if the Web could solve the pet–shelter allocation problem? By midnight, Betsy and Jared made a New Year's resolution to start a new website that would aggregate information from shelters and rescues and put it all online.

Betsy and Jared launched Petfinder just months later. They initially called various New Jersey rescue and shelter groups, which, after some raised eyebrows and skepticism that comes with an out-of-the-box idea, would snail-mail or fax them pictures and information on adoptable and needy dogs and cats to upload to the website. In short order, participating shelters emptied out their inventory of animals, and the field of sheltering would never be the same. It also had the effect of galvanizing caring people—often with a particular interest in a breed or a region but no knowledge that so many Chihuahuas or beagles were in need—to form rescue and fostering groups.

Petfinder.com's popularity quickly outgrew this model, and within two years Betsy and Jared took it national. Today, about 12,000 shelters and rescues upload photos and information—with more than 250,000 animals pictured on the site at any one time—as a way of creating a massive virtual kennel for would-be pet parents. By using the search function, people can locate almost any dog or cat of their liking in their area—finding even the most obscure breeds, and allowing them to skip the step of going to a shelter and coming up empty because of a limited selection. Since it went online, Petfinder.com has done for adoption what Match.com and eHarmony did for dating—allowing suitors to assess key criteria and zero in on the most suitable prospects (the difference is, with adoption, it's a lifetime commitment).

After research on Petfinder.com, your next step is the face-to-snout encounter, allowing you to get the feel of a wet nose or hear a welcome meow. In two decades, Petfinder.com has made possible more than twenty-two million pet adoptions—twice the number of dogs and cats adopted out through PetSmart and Petco. The site now even has adoption information for rabbits, parrots, reptiles, and all the other creatures large and small that shelters and rescues now welcome. It's now becoming broadly disruptive to the other sectors of the pet trade, which have their own ghastly mills for birds, rabbits, hamsters, and other animals.

Through a combination of the online promotions pioneered by Petfinder.com, the use of a PetSmart in-store adoption program, and the creative and dedicated work of Lost Dog and Cat Rescue, my Lily became an adoption success rather than a euthanasia statistic. She was cared for and taken in instead of discarded and forgotten. The goal of everyone involved is to replicate this outcome over and over, so that shelters don't face the horrible task of killing healthy animals for lack of space or money. So many good and resourceful people today are striving to achieve that goal. They're running high-volume spay-and-neuter clinics, lobbying to secure public funding for sterilization and bans on pet stores selling puppy mill dogs, and transporting animals from high-kill areas to high-adoption regions. They're assisting people and their pets in disadvantaged neighborhoods, developing best practices in the field, and advertising to knock down stereotypes about shelter dogs. They're conducting trap-neuter-return programs for community cats as an alternative to sheltering and euthanizing them and even more.

PetSmart and Petco are mixed breeds when it comes to tackling animal homelessness, part for-profit entities and part charitable enterprises. PetSmart Charities has become the biggest foundation for animals, and the Petco Foundation isn't far behind—collectively, they give $70 million every year to programs and other charities to help dogs and cats. Their revenue-generating strategy is simple and effective: asking PetSmart and Petco customers to round up their purchase to help animals. When millions of their customers make small contributions, they add up in a hurry. And that charitable giving is all hitched to the success of PetSmart and Petco as part of the new, humane economy.

They are two among a new generation of well-funded, high-impact charities and foundations. David Duffield, the former CEO of PeopleSoft and cofounder of Workday, has committed more than a billion dollars to his foundation, named in honor of his beloved miniature schnauzer, Maddie. Under the leadership of Duffield and its original CEO Richard Avanzino, Maddie's Fund has embraced community-based solutions to get more animals into loving homes and to achieve "a no-kill nation." The foundation is partnering with The Ad Council and the HSUS to lead a novel nationwide advertising campaign called "The Shelter Pet Project"—so far generating more than a quarter billion dollars' worth of advertising. The campaign reminds people that cats and dogs typically end up in shelters because of the shortcomings of humans and not because of anything the animals did. It was a lost job, a divorce, or some change in fortune that left a good animal in a bad situation or worse.

These are game-changing numbers, and with stronger leaders at the local level we're seeing great outcomes. No community has had a bigger and more inspiring turnaround than Washington, D.C. When Lisa LaFontaine arrived at the Washington Humane Society in 2007 as chief executive, the shelter was struggling with an annual intake of about ten thousand cats and dogs and a deficient facility built fifty years ago. The Society's "save rate" was just 28 percent, meaning that nearly three quarters of the cats and dogs arriving at its New York Avenue facility were not coming out alive. But LaFontaine embarked on a series of reforms: offering dogs and cats at events in the community; setting up a city-wide, low-cost spay-and-neuter program; expanding the foster program and partnering with rescue groups; starting aggressive trap-neuter-return (TNR) for outdoor cats; focusing on customer service; and building public awareness of the scale of the problem. By 2015, the Washington Humane Society's "save rate" had more than tripled to 90 percent, meaning that nearly nine out of ten cats and dogs found new homes. That's the case even though the organization has a disproportionately high intake of pit bulls, making the feat even more remarkable.

More than ever, there's an expansive network of charitable and government enterprises, prepared to deliver on the hope of a humane economy. In a nation with 3,100 counties, there are 3,500 brick-and-mortar shelters, collectively spending more than $2.5 billion a year on solving the problem of animal homelessness. About half of them are private animal shelters run by charities, like your local humane society or SPCA, while the other half are run by local governments and are typically called "animal care and control" facilities. Altogether, these shelters employ a workforce of thirty-five thousand people—an average of ten employees per facility, including administrators, veterinarians, kennel workers, behaviorists, adoption counselors, and others.

There are another 13,000-plus rescue groups. Most are small, with a few if any paid staff and an abundance of volunteers, but some are substantial operations. Lost Dog has an annual budget of more than $1 million, and adopts out 2,500 animals a year. It relies on pet foster parents to increase its capacity and to care for more animals than its shelter can handle while they are being advertised for adoption. Of course, foster homes have a way of becoming forever homes, as many people come to love their foster animals and adopt them. We fondly refer to these cases as "foster failures," but they are really adoption successes.

The countless hours invested by volunteers keep a lid on actual costs and make us all proud of their selflessness. So many men and women feel called to protect animals from neglect, callousness, or cruelty. Naturally, they turn first to the long-known, highly visible problem of homeless pets. But if we could set that problem right once and for all, just think how much good those same people could do in service to equally worthy animal causes—to say nothing of the many other social concerns that could use their compassion and idealism.

My guess is that more than 90 percent of the money going to the cause of animal protection is devoted to addressing the problems of animal homelessness and promoting sterilization of dogs and cats. Animals abused on farms or in testing laboratories, horses re-routed to slaughterhouses, and mistreated wildlife—to name just a few categories of need—all receive a tenth of available resources, even though they represent more than 99 percent of the animals at risk. Imagine if we could simply maintain giving at the current level, but see ten times the money directed at these other problems because we've ended needless euthanasia and brought an end to puppy mills? It would be a new day for the cause of animal protection and bad news for the perpetrators of so many other forms of cruelty.

We've long known that that the best way to get there is to increase adoptions and decrease birth rates for dogs and cats—the strategic pillars of the movement to end euthanasia. Started in 1994 and acquired in 2015 by the American Society for the Prevention of Cruelty to Animals (ASPCA), the Humane Alliance has pioneered a break-even business model to help curb animal homelessness by conducting high-volume sterilization at a low cost. By hiring and training veterinarians to conduct safe and fast sterilizations, and doing marketing and developing partnerships with local animal welfare groups that keep customers streaming in, some facilities are known to conduct as many as twenty-three thousand cat and dog sterilizations every year. The Alliance has now helped 137 start-up spay-and-neuter clinics to replicate its success, collectively spaying and neutering 3.5 million dogs and cats—often in parts of the country that really need it, and keeping things affordable for people and families with modest incomes. Transformative business practices like these, and the forty-year movement to convince Americans that they should spay or neuter their pets, have made all the difference: 83 percent of owned dogs and 91 percent of owned cats are now spayed or neutered in the United States, compared with barely 10 percent in the 1970s.

But sterilization is still a surgical procedure, and it takes time and a licensed veterinarian to perform. In poorer communities, we estimate that 90 percent of owned dogs and cats are reproductively intact, largely because of the cost of spaying and neutering and the scarcity of veterinarians to conduct the procedures. Here the HSUS's Pets for Life program has been game-changing, bringing services and doing outreach to families living at or below the poverty line right in their communities—providing spay-and-neuter vouchers, vaccinations, pet food, and other services and supplies. But what if the HSUS and Lost Dog and other groups could simply give animals an injection, just like the local CVS offers customers a vaccine in advance of flu season? Owners and their pets could then conveniently go in and out of a site at low cost and with no licensed veterinarian required. More than half a century after the birth control pill first became available for women, there's no reason to think that something similar cannot be devised to regulate conception in animals. And if it were affordable and widely available, that treatment would change the equation, especially in underserved communities, and certainly for the tens of millions of reproductively intact feral cats that roam every community in the United States.

It could revolutionize street dog management programs in developing countries, where animal control can be ruthless. Humane Society International (HSI) estimates that nearly half of the world's dogs are roaming the streets—more than 300 million. Imagine. Managing them humanely—instead of clubbing, poisoning, or gassing them, as happens all too frequently in China, Russia, and some other countries—is our next great challenge, as we close in on eliminating the euthanasia of healthy animals in the United States. Rather than capture and shelter and adopt—which is unworkable in nations where there is little or no sheltering capacity and no real tradition of adoptions—our teams now humanely capture the dogs, and sterilize and vaccinate them right on the ground, releasing them immediately. But we can dream of an even better solution.

New technology offers one way to crack the code on animal homelessness. Dr. Gary Michelson, an innovative patent holder and a billionaire who has devoted a large share of his wealth to helping animals, is providing more than $10 million a year through the Found Animals Foundation to develop a single-dose, nonsurgical sterilant for male and female cats and dogs. Researchers have already developed contraceptive vaccines for more than seventy wildlife species. Thus far, though, scientists have been unable to develop a vaccine for dogs and cats because of their species' complex reproductive physiology. Zeuterin, launched in the United States in February 2014, is currently the only nonsurgical sterilant with US Food & Drug Administration approval. It's designed just for male dogs between three and ten months of age, and it's a caustic agent—calcium chloride—that destroys the seminiferous tubules in the testicles, preventing the production of sperm. It's been in the works for years, and researchers are continuing to refine the methods and reduce side effects, which now occur in about 1 percent of cases. Zeuterin is now being used in some shelters and private veterinary clinics. Rose Wilson, who used it at her shelter in Lawton, Oklahoma, in 2014, told the *Wall Street Journal* that the drug is "simple, it's inexpensive, and it's painless. This is the best thing that's happened in the spay/neuter world in a long, long time."

Many people in the field consider Zeuterin an important step for the humane economy, but not yet pivotal. And no matter how many males are sterilized, if a single one goes untreated, he can impregnate any number of females. The Alliance for the Contraception of Cats and Dogs (ACCD), working closely with Michelson and his Found Animals Foundation, is steering a broader range of research toward engineering safe, cost-effective, easy-to-deliver, painless ways to limit dog and cat reproduction. While there's been some very serious science, leaders in the field say we're still years away from a single-dose chemical sterilant for females. ACCD faces not only the hurdles of the challenging reproductive biology of dogs and cats, but also the lengthy and costly approval process of the Food and Drug Administration. When it comes, however, a chemical sterilant will be one of the great innovations in the field of animal welfare, with worldwide applications and dramatic benefits.

While that tool would be the long-awaited magic bullet of sorts, we needn't wait for it to end the crisis of animal homelessness and euthanasia. A grab bag of tools exists now—we only need to pull them out and put them to work. A new day for our companion animals is not only in sight; it's upon us.

Since the mid-1970s, thanks to all of the multiple strategies, the number of dogs and cats euthanized has fallen by 80 percent, from fifteen million a year to three million now. Today, more than ever, we are within striking range of ending the euthanasia of healthy and adoptable dogs and cats. But as an aspiration turns into an expectation, we'll need to find more creative ways to get homeless dogs into loving arms through innovative marketing. The dogs and cats who never make it out of shelters are just as good, as sweet, and as loving as the ones who do. Each time one of them is led down that hall to be euthanized, it's a scene that very few of us could watch without tears. The moral cost is heavy, whether we are paying attention or not. And with so many animals still killed for no defensible reason, who doubts that we can do better?

Puppy Mills and "Old School" Pet Stores

On a recent trip through southern California, I went to visit my friend Christina Morgan, the petite, elegantly dressed founder of Paw Works Adoption Center and Boutique at the Oaks Mall in Ventura County. The store has a welcoming feel, with a stone façade and wide and deep custom couches that you'd see in the television room of a high-end home. It's clean and inviting, its sales team friendly and well trained. Paw Works pulls dogs and cats from an animal shelter that Christina helps to fund in nearby Camarillo, and her store makes them available for adoption. The day I was there, I met Belle, a two-year-old Boston Terrier, who somebody had given away on Craigslist. I held Cecily, a one-year-old Silky Terrier who had been a stray, and I patted Cori, a six-month-old Miniature Pinscher mix who had been surrendered by her owner. I told Christina we got our Lily from a PetSmart at a mall, and observed that she was wise not to simply wait for people to come to her Camarillo shelter.

Community-based adoption programs like Christina's are smart marketing strategies for shelters and rescues. Prospective adopters no longer need to trek to a shelter and steel their spine. So many people simply won't go to shelters because they fear they will find these places sad, uninviting, or inconvenient. In an age of customer service, where you can get money out of a machine and have just about anything delivered to your door, you've got to meet people where they are—whether at a pet store, a shopping mall, a big-box store, a fair, or a park. Just about everybody agrees that euthanizing healthy animals is a serious moral problem, but they're not making the connection that buying a dog from a conventional pet store may perpetuate that problem. They default to the pet store purchase perhaps because it's convenient, or on the false assumption that these outlets sell well-bred animals while the shelters offer only mutts or misfits. Compete with the pet stores on convenience and show off mutts and purebreds from the shelters, and you'll win their hearts in the marketplace. At a shelter with a high save rate, or a rescue group moving a large number of dogs and cats into homes, you'll see savvy marketers and a schedule of community-based adoption events that put dogs and cats in front of people and pull on their heart strings for all the right reasons.

"Our dogs are not only young and generally healthier than the dogs at the pet store, but we have directly intervened and are trying to help animals at risk of euthanasia," Christina told me. "I also offer them at a cheaper price. If people were making logical choices, they'd be coming to my store." That's where the people at Paw Works Adoption Center and other rescue groups come in, providing a safety net and then letting the dogs and cats do their part in winning new hearts and new homes.

After visiting Christina's boutique, I asked her to take me to Barkworks, Pups, and Stuff, a conventional pet store at the far end of that same mall in Thousand Oaks. It took us five minutes of walking past Nordstrom, Coldwater Creek, and Pottery Barn to reach Barkworks, where, of course, the only animals on display were puppies and kittens. The puppies tussled and tumbled over each other in their glass enclosures facing out to the mall's concourse with shredded paper clinging to their heads and tails. They were all fluffy and fetching, full of innocence and playfulness—almost too much to bear for a dog or cat lover. If you didn't know better, you'd reach for your credit card then and there and walk out with a puppy under your arm. You want one, not only because they are irresistibly adorable, but also because you feel a little sorry for them. Every night, when the staff leaves and the mall is closed, they're left all alone in the dark. It's not abuse they're enduring, but it's far from a loving home. And that's precisely the sales formula. People typically buy puppies from pet stores on impulse, motivated by the feelings of warmth and sympathy that the storeowners try very hard to elicit.

At Barkworks, the Bichons, Yorkies, Pomeranians, French bulldogs, and other small purebreds go for $700 to $2,500—as much as ten times Christina's $250 adoption fee for dogs. For customers concerned about puppy mills, Barkworks has signage to assure them that the dogs absolutely don't come from those sorts of places: "We work with only responsible, USDA-approved breeders." What the signage doesn't say is that some of these breeders are anything but "responsible." We've investigated so-called "USDA-approved" mills that supply conventional pet stores, and time and again, the records amount to a rap sheet of animal welfare violations—and even then the government doesn't shut them down and they keep selling puppies. And even if they did comply with the rules, that wouldn't guarantee much of a life for the dogs. The reality is that the USDA rules are more akin to survival standards, with the puppy mills cutting corners, giving dogs barely enough care to keep them alive, and the pet stores, whether unwittingly or not, selling their customers a bill of goods.

The market for pets has always functioned in an imperfect way—in fact, it's upside-down. For a dog from a mill, customers pay many times what it would cost to get one from a shelter. They assume that the higher price means better quality—good genetics, sound behavior, and all the benefits of animals who've been given the necessary medical treatments. In fact, exactly the opposite is true.

At the HSUS, we learn of hundreds of cases a year of people buying puppies from pet stores only to rack up veterinary bills in the thousands of dollars, or, worse, to watch their new puppy die. Puppy mill dogs are not sterilized or vaccinated, and almost never see a veterinarian. But the biggest hidden costs for purchasers of puppy mill dogs are the genetic and hereditary problems of these purebreds—which the mills turn out indiscriminately. Many purebreds face chronic pain and shortened lifespans—facing maladies, depending on the breed, from heart problems to hip dysplasia to certain forms of cancer. It's not just that the mill dogs are purebreds (a third of dogs in shelters are purebreds too), but that they are often severely inbred over many generations, magnifying the hereditary problems. I've heard it again and again from veterinarians: you so often do better when you get a dog from a shelter or rescue.

A rescue dog is also a better financial bet because rescues and shelters front-load animal care services, with the adoption fee covering veterinary care, vaccinations, and sterilization. A $250 adoption fee is hardly a freebie, but even the least expensive dogs at a pet store go for more than that. An adopter gets a bargain-basement rate because volunteers cover most of the labor costs while donations to the shelter or rescue cover the food and veterinary expenses. And there's no profit margin built into the enterprise because it is charitable. The exception is when you take on a hardship case as we did with Lily. Because she had obviously been neglected and needed special treatment, Lost Dog transferred some medical costs to her new caretakers—Lisa and me. But as money goes, we've never gotten so much out of so little.

With Christina Morgan and other rescuers like her working hard to adopt animals in need, why should she have to compete with the very people, in the mills and the pet stores, who are causing the problem of overpopulation? At the Oaks Mall, Barkworks is drawing away some of the same customers and families who could be adopting more of Christina's rescue dogs. The store is only generating revenue for its owners and staff and for the mills, which might be a thousand or more miles away. And because Barkworks is displacing animals in the adoption pipeline locally, it's imposing a burden on its community, where many animal shelters suffer from overcrowding and underfunding.

Think of my little Lily, who may herself have come from a puppy mill and at some time was abandoned. The Spotsylvania county shelter may have picked her up on the street, fronting the cost of maintaining trucks and personnel who gather up wayward strays. And even if she was relinquished directly to the shelter, the facility absorbed the costs of feeding and caring for her. The shelter was about to face the cost of euthanasia and disposal too—a cost not just measured in dollars, but in the daily grief of shelter workers, people who set out to help animals but instead find themselves killing animals as a matter of routine. Thankfully, Lost Dog intervened. When they pulled Lily out and gave her a second chance, this group then picked up the costs of her care, transportation, and shelter. Veterinary exams, medicines, and the spay surgery cost hundreds of dollars. Lost Dog held her for weeks as she went to nine adoption events until she found a home. Lisa and I then took over, paying a $375 adoption fee to Lost Dog—a substantial though incomplete offset of their costs (if a rescue tried to recoup all its costs through the adoption fee, that price tag would deter adoptions). On the day of adoption, Lisa and I rang up a bill of more than $400 at PetSmart, picking up toys, beds, pet food, collars, leashes, and other stuff for Lily.

After allowing Lily to adjust to her new environment and to us, we went to a veterinarian, and began over the next few months to treat her problems. We medicated her for Lyme disease and parasites, treated her ear and eye infections, got her teeth cleaned and the broken ones pulled, and tried a battery of treatments to stop her from gnawing on her left foreleg. The veterinarian even prescribed antidepressants for her. In all, we spent more than $2,000 to help her during her first year with us. This was money we had and that we paid willingly. We didn't know at the beginning about all of her ailments, but we knew she was a hardship case—a case of neglect and perhaps even cruelty, but one we embraced as an opportunity to make her whole. We cannot help but think of animals who aren't so lucky.

Our costs were mitigated somewhat because right after we brought Lily into our lives, we also purchased PetPlan insurance. Only 1 percent of American pets have healthcare plans, compared to 50 percent in Sweden. But by paying a $22 a month premium, we saved more than $1,000 on medical costs. With rapidly improving veterinary medicine that allows pets to live longer lives, and specialists whose expertise tracks that of human healthcare fields, it just makes sense to rely on insurance as a buffer against a catastrophic incident or some other major medical expense. No responsible pet owner or guardian wants to forgo care for their animals because they can't afford it. As with human healthcare, the costs of veterinary care have been surging, and market-based pricing is beyond the reach of so many pet-keeping families.

The point is that all of these costs, to help one animal in need, could have been avoided were it not for cruel and careless people who created that need in the first place. As is so often the case with industries that abuse animals, the true costs are largely externalized. Their savings, in discarding one used-up dog, merely imposed an investment of resources many times over to undo the consequences of their thoughtlessness and greed. As an economic model, it sure works for them. The only downside is, it's a net loss for everyone that is left to pick up the costs as long as such practices are permitted.

The story of one other dog in Virginia particularly resonated with me because it brought to mind so many aspects of Lily's story. Bailey was also a beagle, a common breed for mills because these dogs are small and trusting. Also perhaps like Lily, too, Bailey found her forever home after coming out of a pet store in Fairfax County. Without realizing the consequences, Felicia Collins Ocumarez had purchased Bailey at the Petland there, instead of going down the road and adopting a beagle through PetSmart. And Bailey also had a host of health problems brought on by stress, inattention, and poor living conditions at the mill she came from. Collins Ocumarez says Bailey was emaciated and overrun with fleas when she brought her home from Petland: "She was under three pounds . . . infested with worms, parasites . . . respiratory issues." All of this cost Collins Ocumarez hundreds of dollars in care—on top of the $1,300 she paid at the cash register for Bailey. Her story inspired a namesake law in Virginia, "Bailey's Law," which now requires pet stores to post information on a dog's origins, so that owners can check the USDA records of the kennel where the puppy originated. Petland didn't pay any of Bailey's bills, leaving a trail of costs for Collins Ocumarez or some other would-be pet owner anxious to bond with a little, loving companion.

Petland, a national chain of stores that lags behind PetSmart and Petco in its growth and success, has been heavily criticized by the humane community for procuring dogs from puppy mills and passing them on to consumers. It has refused to stop selling dogs from puppy mills. Through the years, it's been the focus of pickets, press exposés, investigations, and other headaches that could not have helped its bottom line. Since the HSUS brought attention to the company's reliance on puppy mills, more than thirty of its one hundred or so stores have shut down. We talk frequently to the company's leadership, and, to their credit, they're experimenting with adoption programs at a few stores and seeking to create humane breeding standards for the dogs they intend to sell. But many franchise owners still cling to the notion that they need to sell dogs and cats to make their business work, no matter the cost. That's outdated economic thinking that customers can bring to an end—by simply taking their business elsewhere.

Petland cannot claim that only PetSmart and Petco can make the new humane pet store model work. In 2014, Pets Plus Natural, with ten stores in the Philadelphia area, adopted a "puppy friendly" policy to replace puppy sales with adoptions from local shelters and "high-kill" shelters in the South. After converting its first few stores, within four months, it had adopted 1,200 dogs to customers—cutting off a sales outlet for mills and expanding the adoption network for shelters. And when it did the right thing, Pets Plus Natural helped animals and won the loyalty of rescue groups and adopters. In turn, they became the boosters and ambassadors of the Pets Plus Natural brand.

This is a process that's playing out store by store in communities throughout the nation, and we're the first to heap praise on pet stores severing their connections with puppy mills and introducing themselves in a new way to the local shelters and rescues. That momentum can only pick up as public opinion, competitors, and political leaders demand a new course. Chicago, Los Angeles, Miami, New York, Phoenix, and more than 100 other big cities now restrict selling dogs from puppy mills. When elected officials, responding to citizens' concerns, do the economic and moral analysis, you know the puppy mills and their distributors are operating on borrowed time.

When I speak to people, however, I urge them not to use this inevitability as an excuse for inaction. We all need to encourage businesses to make the right decisions, and to reward them for doing so. We need to help our friends, families, and neighbors make wise choices in a domain where emotion can sometimes get the best of us. Together, we can hasten the day when puppy mills are a thing of the past, and every pet store is a place of redemption and hope for an animal in need.

Breaking the Chain of Dogfighting

When the lead member of the SWAT team made his first few attempts to break down the front door, he quickly realized something more than a firm kick was in order. Robin Stinson, a convicted felon and a suspected dogfighter, had reinforced the doors with steel in his refurbished double-wide, green-paneled white trailer, and then dropped wooden planks across the doors to impede forced entry. Apparently, he suspected he might draw some uninvited guests, whether angry dogfighters, rival drug dealers, or perhaps even a SWAT team like the one massed outside his home that morning. The man had good reason to be concerned about the last group, since the county sheriff's office, acting on a tip, had obtained a search warrant two years earlier and turned his trailer inside out. Deputies didn't uncover enough evidence to bring charges then, but they had a feeling there would be another visit.

Unbowed, or maybe just unaware of law enforcement's continuing interest in him, Stinson persisted in fighting dogs at pits throughout the South, and continued to keep dogs on his property. On the morning we showed up alongside the sheriff's deputies, eighteen pit bulls strained on six-foot-long metal chains staked into the dirt. Most of the dogs had a small rudimentary wooden enclosure, providing just the barest relief from southern Alabama's late August sun.

I had set my alarm for 3 a.m., as had about fifty other HSUS personnel and volunteers assigned to this site and six others for possible raids in Alabama and southern Georgia. Our team met downstairs at 3:30 a.m. at a Hampton Inn in Dothan, all of us in our blue Animal Rescue Team jackets. We drove about twenty minutes to the outskirts of town to an old peanut warehouse that now serves as a staging area for the Houston County Sheriff's office. We waited for clearance to proceed to the site an hour away, heading through the early morning fog as we passed over the Pea River and into Elba until we found Stinson's home.

The fortified door proved only a temporary problem. A battering ram would do. While some members of the SWAT team were knocking down the door, others broke through a large-frame window, throwing flash grenades into the bedroom to disorient the suspect as the teams rushed in behind. Stinson was thought to be armed and dangerous.

He started to run, but with a dozen officers drawing weapons he quickly figured out the situation and surrendered. By 6:30 a.m., he was in the custody of US Marshalls and on his way to Montgomery for booking.

After police secured the premises, our HSUS team moved in to handle and seize the dogs. I mapped the property. At the center was the house trailer, with its large wrap-around patio. A cedar fence extended almost to the outer reaches of the property. In the yard were an aboveground swimming pool, a satellite dish, and scattered mounds of junk—rusting chairs, stacks of wooden pylons, and metal pipes. According to the sheriff's deputies who had been on the site two years earlier, Stinson had recently upgraded his home. The trailer had doubled in size, with six freshly painted bedrooms. The fencing was all new. It wasn't opulent, but it wasn't too bad for a guy without a regular job.

The chained dogs were at the rear of the property, where the fence made them invisible from the road. The chains were not long enough for any of the dogs to interact, and that was by design—after all, they had been bred and trained to fight. The fencing wasn't quite finished, so you could see some of the tethered dogs from the road if you were in just the right spot. Two little dogs were kept in rabbit hutches with feces piled up beneath them.

Despite years of neglect and suffering in such disgraceful conditions, the dogs greeted us with ears down and wiggling rear ends and tails. A few of them were hyper, starved for attention after having been denied affection for so long. Seeing the men and women who had come to rescue them, they seemed to instantly know that here was a different breed of person, clearly unlike what they were used to. Every one of them welcomed clean water, since the metal bowls within reach of their chains contained two inches of stagnant, fetid, discolored slurry. The dogs couldn't dispense enough kisses or get close enough to the volunteers. The ones I bent down to pat wagged their tails; even their tongues seemed to be wagging too. My colleagues all got into the act, taking a little time to comfort the animals before we started documenting the scene, taking pictures of the dogs, and then delivering each one to our veterinary teams stationed out front.

We didn't need a diagnosis from the veterinarians to conclude that many of the dogs were undernourished. The older dogs had scars on their faces and forelimbs. Fleas covered other body parts. One sickly looking, emaciated older dog had been eating his own vomit. Another had to be rushed directly to a veterinary hospital.

My colleagues joined law enforcement agents at six other locations that morning for arrests and dog seizures. All told, at the thirteen sites raided by police, the HSUS, the ASPCA, and other animal welfare groups rescued 367 dogs. It is thought to be the second-largest dogfighting bust ever in the United States. (The biggest one was an eight-state operation centered in Missouri in July 2009, where we joined law enforcement teams and other animal groups in seizing more than five hundred dogs.) By the time every lead had been chased down, using evidence gathered in the initial raids, we'd helped to rescue more than four hundred dogs.

Ten suspects, including Stinson, were arrested and indicted on felony dogfighting charges. Federal and local officials also seized firearms and drugs, as well as more than $500,000 in gambling proceeds. And they discovered the remains of dead animals on some properties where dogs were housed and allegedly fought. "The lowest places in hell would be reserved for those who would commit cruelty to animals," said George L. Beck, Jr., US Attorney for the Middle District of Alabama, when he announced the indictments.

Throughout that morning, my mind raced with thoughts. If the dogs weren't too damaged, we could turn their lives around with veterinary attention and behavioral work, and get them rehabilitated and then adopted. I also reflected on how quickly fortunes could change, even in the lonely life of an abused dog. They had all been destined for life at the end of a heavy chain, untethered only when they entered a dogfighting pit. But thanks to the SWAT teams and our Animal Rescue Team, they'd experienced, perhaps for the first time, sustained human kindness. At the very least, they'd never see the inside of a dogfighting pit again.

In a different way, this operation would also turn around the lives of the dogfighting suspects. Perhaps these men had believed they were beyond the reach of the law and that they would suffer no consequences from profiting from animal cruelty. After they had chained and penned up the animals, now the state had a pen of its own for them. Even when their families or friends visited, there would be a wall between them. Was the dogfighting business so lucrative and exhilarating that Stinson and other suspects would risk their freedom for it? Would a taste of confinement make them think about the far worse punishments they had inflicted on those dogs?

Dogfighting is tawdry and crude up close, but it is also a big, global business. And most of the bad actors involved in it don't have to worry about arrest. It's still legal in more than 120 countries. And even where dogfighting is banned, there are tens of thousands of dogs caught up in the enterprise and billions wagered on the merciless spectacle of the fight. It's an example of the old, inhumane economy at work, and we're not trying to make it better—we're battling to eradicate it entirely, to rid every society on earth of this form of barbarism. That's why we've campaigned to make dogfighting a felony in every state and succeeded in making it the most severely criminalized form of animal abuse of any type.

The toll of this enterprise, as with puppy mills and pet homelessness, is inflicted foremost on innocent animals. But there are many indirect costs as well, including the heavy costs to charitable animal organizations and to agencies of government that must respond and care for the animals. After the raid of Stinson's property with the SWAT teams, animal welfare groups had to hold almost four hundred dogs for evidentiary purposes, while nursing them back to health and re-socializing them for life in a home. Just HSUS's share of the cost, accrued over more than a year's time, was $2.5 million to care for these dogs. Multiply that by the hundreds of dogfighting busts that the HSUS and other organizations conduct, and the price tag is in the tens of millions annually. Add in the seizures for hoarding cases, other forms of cruelty or neglect, and puppy mills, and the price tag starts to look astronomical, at least for charities with plenty of other crises to manage.

It is truly galling that the people contributing to the problem of animal homelessness—the dogfighters, puppy millers, and pet stores that traffic in these animals—impose such enormous costs on others. The sick, overbred, injured, and cast-off animals are only their first victims. So often, these industries and abusers—with their hack public relations hirelings, trade associations, and networks of apologists—proclaim the right to do whatever they want with animals. The puppy mill operators even sued the USDA to stop it from regulating Internet sellers of puppies. Yet, when our HSUS Animal Rescue Team raids one of these operations, it may cost us more than $100,000 to deploy, care for the animals' medical needs, provide emergency shelter and transportation, and then work with our emergency placement partners to get the animals adopted (the costs for the Alabama dogfighting case were inordinately large). That's the hard-earned money of HSUS members, spent to right the wrongs of people who keep causing these problems and couldn't care less about the consequences. If only the law would finally put an end to the whole filthy business, eliminating puppy mills entirely, we and other shelters and rescue groups could do a lot more good on other fronts.

Toward the end of 2014, a federal judge meted out the most severe penalties for anyone ever involved in this enterprise. One ringleader of the operation, a fifty-year-old man named Donnie Anderson, of Auburn, received an eight-year sentence after pleading guilty to conspiracy, sponsoring dogfights, possessing a fighting dog, and operating an illegal gambling business. Several others, including Robin Stinson, also received stern sentences.

When you're involved in animal rescue work, you have to marvel sometimes at all of the planning, effort, and sheer human energy that some people devote to doing cruel things. Sad as it is to consider, there are men and women out there for whom dogfighting spectacles are a passion, the closest thing they know to a purpose-driven life. And from my vantage point, it's only more amazing to look around every day and see all the people who live for a very difference purpose—helping animals instead of hurting them. It takes a lot of idealism, energy, and also resourcefulness, and it is just such people who give dogs like my Lily a story with a happy ending.

CHAPTER TWO

Big Ag Gets Its Hen House in Order

Man must pass from old to new,
From what once seemed good, to what now
Proves best;
How could man have progression otherwise?

—ROBERT BROWNING

A Capitalist Revolution Frees the Pigs

"Yeah, hi, Wayne, it's Carl Icahn. I understand you are working against animal cruelty, and I'd like to help." Though I've come to expect that this cause can attract the unlikeliest of supporters, I was still pretty surprised on that day in 2011 to hear from one of the toughest guys on Wall Street. Business writers typically refer to the legendary Icahn as an "activist investor," though his "activism" is usually reported to focus on securing board seats, conducting proxy fights, and dumping management teams to drive up share prices and make money—for Icahn Enterprises and his investors. Social responsibility is not exactly thought to be his guiding inspiration. A Princeton graduate, nearing age eighty, with a voice both resonant and raspy and experience and wealth that give him staying power, he's a master at buying a large stake in a company, often from a hostile position, and bending it to his will. With assets estimated at $25 billion, Icahn is not a man, at least in matters of business, to be either trifled with or intimidated.

We at the HSUS count powerful supporters within our ranks—in politics, business, the arts, and elsewhere. But here's someone able to command the attention of the world's most powerful CEOs, out of a mix of fear and respect. While animal-protection advocates have become expert at protests and letter-writing campaigns, and skilled at public policy and courtroom maneuverings, the domain of corporate deal-making and leverage is its own beast altogether.

After taking Icahn's encouraging phone call, and after meeting with him a couple of times at his office in the General Motors Building in midtown Manhattan, I decided to aim high. I reminded him that of all animals at risk from human conduct, the greatest numbers are caught up in our increasingly harsh, detached industrial food production system. When some people who are unfamiliar with our work hear the words, "Humane Society," they may think of only pets or horses, but we've never veered from our mission of protecting all animals, or from taking on the toughest fights. To address factory farming, I asked Icahn for help with our efforts to change McDonald's animal welfare policies, knowing that the public supported our goals but the company was an exceedingly powerful one that might need an extra kick in the pants that only he could deliver.

"We are trying to convince the biggest food companies to institute changes in their purchasing practices," I explained, "and few corporations have a bigger impact on the supply chain and animal agriculture than McDonald's." The company buys about 1 percent of US pork, but about 15 percent of all US pork bellies, which means it takes a piece of one of every seven pigs in the nation. If McDonald's required its suppliers to observe higher animal care standards, that decision would reverberate throughout the pig industry and perhaps even the entire food sector.

The vast majority of breeding pigs in commercial production, I explained to Icahn, are kept in two-foot-by-seven-foot cages called gestation crates—about as big as a coffin, but confining animals still very much alive and about twice the weight of a full-grown man. "Forget the pig is an animal," wrote an editor in *Hog Farm Management* as far back as 1976, when the era of crate confinement was ascendant. "Treat him just like a machine in a factory." That attitude came to govern the industry, with vertically integrated, industrialized operations gradually coming to dominate production, and forcing hundreds of thousands of family farmers out of business. In fact, in 1980, there were nearly 700,000 pig farmers, and today there are fewer than 60,000, even as the number of pigs raised for food has climbed substantially during the same period. Sow confinement in crates is so severe that the animals cannot walk, turn around, or play with other animals. They are held in place for years on end—except when they are briefly moved to give birth and held in equally confining farrowing crates. They chew the bars of their crate till their teeth crack, or bang their foreheads and scrape their snouts so violently they bloody them. Victims of solitary confinement, they may go mad—a punishing psychological experience to match their physical bruising.

"That's unconscionable and inhumane," Icahn said in response, reiterating his offer to help. "Even if a CEO is on a golf course," he told me, "he'll generally get back to me quick if I put a call in."

We had been dealing with McDonald's for years, but had stepped up our campaign in the months before Icahn's call came in out of the blue. Our team had been in regular contact with Bob Langert, then vice president for sustainability at McDonald's, but the negotiations and the progress had been slow going. We learned in the process that Langert, for all of his abilities, was not the final decision maker. People pushing a wide variety of social causes came his way every day—pleading their case on obesity, packaging and waste, the minimum wage, GMOs, trans fats, and palm oil, among other subjects. We aren't the only cause that knows that the pathway to progress is through changing the supply chain decisions of America's biggest food buyers.

Icahn suggested that breaking through to the senior ranks of McDonald's decision makers could best be accomplished by nominating a candidate for the company's board of directors. He warned that forcibly winning a board seat would be very difficult, given the company's market capitalization of $100 billion. Nevertheless, he urged me to seek a board seat at McDonald's. With "billions and billions" of hamburgers served, as the signs below the golden arches proclaimed, the company had run through countless millions of animals—and that process of buying up vast numbers of animals to serve as Big Macs, Sausage McMuffins, and Chicken McNuggets would only continue via twenty-five thousand outlets worldwide. Why shouldn't a company with commerce centered on animals have a guy with a special focus on animal welfare on the board? Icahn thought it would help the company's credibility given the criticisms about its business practices and the growing public debate about the treatment of animals. It made sense to me, too, but I thought that it would be a nonstarter with the company brass.

Icahn placed a call to the president, leaving a message with his assistant. Even for a company known for quick service, it was telling that Don Thompson got back to Icahn in no time at all. The three of us jumped on the line. Almost effortlessly, Icahn had elevated the discussion of animal welfare from the office of social responsibility to the C-Suite. Icahn told Thompson that confining mother pigs in crates was unacceptable and, moreover, that he wanted me on the board. Thompson was unfailingly polite and said McDonald's was committed to doing the right thing on animal welfare. He suggested we talk through the crate issue, and deftly deflected the board selection idea. He noted that Langert had been keeping him informed and emphasized that the company, with its enormous supply chain and millions of customers, had to be practical. McDonald's also had thousands of franchisees, and relationships with meat packers and farmers, and couldn't impose standards from on high without involving these stakeholders. Even so, Thompson insisted McDonald's wanted to do better, noting that years earlier it had established more humane slaughter standards and required egg suppliers to offer slightly more cage space for hens. It wasn't a "yes," but it also wasn't a "no"—and I was really starting to like our new lead negotiator.

As discussions continued, with Icahn and me on a series of calls with Thompson, I wondered what the executives at McDonald's must have been thinking. I imagined that they thought, "If we don't do something, Icahn is just brash enough to make our lives difficult and to push this onto the front page of the *Wall Street Journal.*" For our part, Icahn and I had agreed to work with McDonald's as constructively as we could. We made the case that public attitudes were shifting. We cited the humane movement's wins on anti-crate ballot measures in Arizona (2006), California (2008), and Florida (2002); the European Union's ban on crates (1999); and new research that showed that group housing systems were better for sows and competitive on cost with crates. One enormous factor was that a major McDonald's supplier had already pivoted: Smithfield Foods had agreed, in the wake of our Arizona ballot win, to phase out the crates. With a million sows in crates, Smithfield was the biggest pig producer in the world, and for years, the most notorious and arguably the most resistant to ending the harshest practices, such as confinement in gestation crates. If an operator as massive as Smithfield could do a turnaround and commit to a phase-out of crates, then why couldn't other producers too? A company with as much capital invested in housing systems as Smithfield was not going to leap without looking. Having once defended crates as essential, Smithfield had turned a corner and determined that alternative methods not only were viable and cost effective, they were the future. Here, in steady, meaningful steps, was the humane economy at work.

More back-and-forth discussions and some brinkmanship ensued on both sides, but eventually Bob Langert called me and my colleague, Paul Shapiro, and Don Thompson called Carl Icahn to tell us that they would get crates out of their supply chain. Langert told us they'd make a joint announcement with us—but they let us know they'd rather not talk at the time about Icahn's role (this is the first public report of his involvement). McDonald's also said it needed a phase-in period—about a decade, in fact—to completely get rid of the crates, largely because most producers had too much invested in their confinement systems and would only swap out their old crates as they depreciated. Indeed, in prior years, farmers had invested billions of dollars in the crates, and it would be economically unworkable, they asserted, to eliminate them overnight. "McDonald's believes gestation stalls are not a sustainable production system for the future," the company announced in February 2012 in a landmark release with the HSUS. "There are alternatives that we think are better for the welfare of sows. McDonald's wants to see the end of sow confinement in gestation stalls in our supply chain."

Some pork industry "leaders" were livid, but they bit their tongues and kept their grousing within their fraternity and outside of the press. They recognized the groundbreaking change for what it was—the giant retailer's acknowledgment that consumers were owed a voice in how animals were treated in the food system. "This is a business decision that they made," Dave Warner, spokesman for the National Pork Producers Council (NPPC), told the press, "We have no problem with that. It is the free market, and it's not the federal government dictating." That acknowledgment did not, however, stop these industry laggards from continuing with tired old arguments for the status quo. Apparently, industry knew best, and consumers should eat their sandwiches and pork chops and pipe down. Yielding to consumers was a dangerous idea. They knew that other retailers would feel heightened pressure to change, and the NPPC took the posture that all housing systems have their advantages and drawbacks—a conveniently ambiguous position for defending the status quo. "So our animals can't turn around for the two-and-a-half years that they are in stalls producing piglets," Warner told the press just a couple of months later. "I don't know who asked the sow if she wanted to turn around." Even with one of its biggest customers renouncing the crates, the pork industry was not backing down.

Yet, as Warner acknowledged, this was the marketplace at work and the McDonald's announcement was a breakthrough, a triumph of public sensibilities over corporate rigidity. One of the biggest food sellers in the world had said it needed pork, but only if the sows had not first been jammed in small metal stalls. I never got my board seat with McDonald's: too bad, since that would have been a stronger statement of leadership still. But the dimensions of the debate, and the sows' housing area, would be forever changed.

Our HSUS Farm Animal Protection division—already a formidable group with great contacts within the food industry—learned a trick or two from Icahn along the way, and before long the campaign against gestation crates would turn into something of a rout. Our team, led by Paul Shapiro, Josh Balk, and Matt Prescott, invited the food procurement people from the big food retailers to two- or three-day retreats, with problem solving sessions, social time, and talk of research results and the future of food and retail. Many of the top officials of these companies, good people through and through, developed friendships with our folks. They had all inherited their procurement systems and, once they learned the details of industrial agriculture, wanted to do their part to make things better. Our staff members equipped them with the best practical and economic arguments, so that they could perform like well-prepared trial lawyers making their best case to a jury of their peers in their executive offices.

The results took all of us by surprise. Other brands rushed to fall in line, and we relished it. Burger King, which had previously committed to purchase cage-free eggs, did the same on gestation crates just weeks after McDonald's did. So did the other huge names in fast food, including Wendy's, Hardee's, and Carl's Jr. Along the way we got the largest supermarkets, including Safeway and Kroger; the big-box stores, such as Costco and Target; and the biggest restaurant chains—from Appleby's to Cracker Barrel to Bob Evans to do the same. The major food service companies—Aramark, Sodexo, and Compass Group, which handle tens of thousands of corporate cafeterias, hospitals, ballparks, and other venues—became the most enthusiastic and highly conscious of food buyers, making pledges not just on crates but on other animal welfare concerns. In the three years after McDonald's pledge, more than sixty major brands agreed to phase out purchasing pork from operations that confine sows so severely.

Farm animal-protection work had, for years, been a slog—an obviously important focus but immensely challenging because of the big, immovable companies on the production and retail sides. The first state to adopt an anti-confinement law for farm animals didn't come until 2002. Before that, the very idea of protecting farm animals was not even a serious-minded subject of focus for plenty of politicians and food retail giants, and the producers were downright hostile. Even though the notion of giving farm animals enough space to turn around seems like the most modest proposition, and the outcome of a campaign to ban the most extreme forms of confinement a no-brainer, that was not always the case. Just a few years before the McDonald's declaration, it was almost unimaginable to those of us in the trenches that dozens of the biggest names would sign up for reform.

We at the HSUS closed the deal, but the most persuasive factor for these companies was the detailed, incontrovertible evidence that consumers were aligned with our values. In fact, consumers reacted with contempt when they began to understand the abject cruelty of industrial livestock farming. The best historical example of that came in the late 1980s, when images of forlorn-looking calves in crates, kept in an intentional state of anemia to make veal, took hold in people's minds and turned them against that industry. Even in my hometown of New Haven, with its famed Little Italy section and veal parmigiana (a cherished local favorite), there was a sense that treating an animal this way came at too high a moral price, and veal vanished from a lot of menus. From 1944 to the late 1980s, American per capita consumption of veal dropped from 8.6 pounds to just 0.3 pounds. The veal industry trade group, seeing that producers had lost the fight over confinement of calves in twenty-two-inch-wide crates, finally pledged in 2007 to nix the stalls and to make the transition to group housing throughout the United States by 2017. Today, there's no discernible rebound in veal-consumption rates, revealing that brand damage can be lasting if an industry accused of cruelty stonewalls and allows negative public impressions to set in. It took two decades and an assortment of tactics—public education and advertising, investigations, lobbying campaigns—but it was possible to shift public tastes and drive a change in consciousness that would forever alter the fortunes of an industry.

In May 2015, HSUS all but clinched the case against the extreme confinement of gestation crates by securing a broad-gauged commitment from the biggest corporate holdout of them all: Walmart, the world's largest food seller, accounting for 25 percent of all grocery sales in the United States. The biggest farms throughout the United States relied directly on Walmart for space in its meat case. Under former CEO Lee Scott's leadership, the company's directives to suppliers about recycling, packaging reduction, and more efficient light bulbs and microwaves had turned Walmart from a laggard into a leader on environmental issues and reduced its energy use and greenhouse gas emissions. With its enormous reach with consumers, it had demonstrated it could move boldly and on a larger scale than even the federal government.

As with McDonald's, however, the step from environmental consciousness to animal welfare at Walmart had not been quick or easy.

My colleagues and I had been in discussions with the company for eight years. I had trekked down to Bentonville, Arkansas, multiple times to meet with the head of the company's global food operations and other key personnel, urging Walmart to show the same leadership and game-changing impact on animal welfare as it had on the environment. So had my colleagues Josh Balk and Paul Shapiro, who had been working with a half dozen Walmart executives and managers, informing them as, one by one, every competitor in the marketplace made animal welfare pledges. On one trip to Bentonville, I was joined by Bill Nicholson, a friend of Lee Scott and himself the former head of the consumer products giant Amway, where he had abruptly ended animal testing after realizing that it was no longer necessary and a drag on his company's brand. We urged Walmart to think big and consider all animals in its supply chain, not just those in crates, and in the end the company did exactly that with this declaration in May 2015:

> "There is growing public interest in how food is produced, and consumers have questions about whether current practices match their values and expectations about the well-being of farm animals. Animal science plays a central role in guiding these practices, but does not always provide clear direction. Increasingly, animal welfare decisions are being considered through a combination of science and ethics."

The company said it was "committed to continuous improvement in the welfare of farm animals in our supply chain," offering the "Five Freedoms" of animal welfare as an ethical framework: freedom from hunger and thirst; freedom from discomfort; freedom from pain, injury, and disease; freedom to express natural behavior; and freedom from fear and distress.

This was an extraordinary advance for Walmart, because the Five Freedoms would apply to all farm animal species in its supply chain.

The policy, given its breadth, would cover not just extreme confinement of pigs and other animals, but also mutilation procedures like tail docking, transport, slaughter, and even the use of antibiotics. Walmart had not been a first adopter, but its scale of operations and comprehensive policy created a new benchmark for just about every farm and food retailer in the nation. We not only had the pork industry on the run over its use of gestation crates, but we got a boost in all of the other areas of concern to us in industrial animal agriculture.

Walmart and McDonald's have become the biggest brand names in food retail by offering low cost as their signature with customers. With the huge volumes of product they move to market at cheap prices, their acceptance of animal welfare standards was a particular triumph for us. If they could do it, other big-scale retailers could, too. Now it was no longer a debate about whether animal welfare was important, but rather about when widely accepted values would be applied. These new corporate policies could not help but trigger a transformation in animal agriculture. It would be much harder now for any company or farmer to sidestep the question of humane treatment by saying that practical necessities trumped ethical concerns. With the public pledges of McDonald's and Walmart, along with the other companies that got on board, we could see a roadmap to the end of the era of intensive confinement of animals on factory farms.

Chipotle, which describes itself as a fast-food company with a conscience, made its stand against extreme confinement of animals a centerpiece of its brand years ago, and it's been building on that notion ever since—with phenomenal success. In 2000, company founder Steve Ells was visiting a hog farm in Thornton, Iowa, where he saw free-range pigs foraging, grunting, and socializing outside, and he was shocked to learn that most pigs, cooped up in factory farms, would never enjoy such simple pleasures. "I didn't want my success or Chipotle's to be based on that," said Ells. So he made a commitment to buy all of Chipotle's pork from farms that do not crate pigs—part of a broader "Food with Integrity" commitment. Chipotle has since campaigned aggressively against factory farming, producing a series of viral YouTube videos that earned millions of views and shined a light on the excesses of industrial agriculture. Ells' decision—benefitting animals but also a boon to small farmers and the environment—has paid off handsomely for the company. Between 1998 and 2013, Chipotle grew from just eighteen restaurants to more than two thousand. Its revenues for the first time ever exceeded a billion dollars in the first quarter of 2015.

That incredible revenue growth—Chipotle had done $727 million in sales in the first quarter just two years earlier—occurred at a time when the company had a major disruption in its supply chain. In January 2015, Chipotle learned that one of its suppliers had violated the company's animal care standards. Chipotle promptly cut off the supplier even though it would face a shortage of higher-welfare pork. Chipotle stopped selling its popular carnitas at more than 1,500 of its restaurants. "We would rather not serve pork at all than serve pork from animals that are raised this way," Chris Arnold, a Chipotle spokesperson told the press. "Replacing the supply we have lost in these ways will take some time, but it is important to us to maintain our high standards for pork, and we will continue to see some shortage while we work to increase the available supply." Food industry analysts noted that the decision "shines their halo" and reminds customers that the company stands for something meaningful, which also burnishes the brand and grows revenues.

Whole Foods Market co-founder John Mackey, co-author with Raj Sisodia of *Conscious Capitalism* and also a board member with HSUS, argues that every company must have a higher purpose than profit making. With his co-CEO Walter Robb, Mackey has reformed not only his company's pork procurement practices, but also its approach to every other kind of animal product sold in its stores. Mackey made

Whole Foods the first major food retailer to adopt a program that raises animal welfare standards and gives customers information to act on their beliefs about any animal product. He provided the inspiration for the Global Animal Partnership (GAP), an animal products certification program with a five-step animal welfare rating system that helps farmers move away from confinement and toward more humane, pasture-based systems.

While other certification systems declare a product humane or not humane—a binary approach—the beauty of GAP's framework is that the rating system allows consumers and producers to ask more of themselves and to embark on a sort of moral climbing expedition. With competition spurring them on, along with the moral adrenaline that comes with doing the right thing, many producers who are already part of the program incorporate more animal welfare standards into their husbandry practices and go from step one to step three and eventually to step five. Consumers, driven by the same sort of moral positive feedback systems, are spurred to pay a bit more and bring home higher-welfare products. With a race to the top on both the supply and demand sides, animals benefit.

By the end of 2015, GAP was connecting millions of consumers with more than three thousand farms through more than four hundred Whole Foods stores, and was the most important and comprehensive animal welfare program associated with agriculture. Overall, Whole Foods does $16 billion in annual sales, and reports that sales of more ethically sourced animal products are robust and growing. Other major food retailers have mimicked Whole Foods' in-store innovations, and I expect other industry leaders to adopt the GAP program or develop their own animal welfare standards over time. Even with the other giants in food retail not yet on board, and Whole Foods as the biggest purchaser of GAP-certified products, there are now more than 300 million animals certified under the program and free from the harshest forms of confinement on factory farms.

The McDonald's announcement, even with the intervention of Carl Icahn, could never have happened without the right social and economic climate. That climate was fostered by the pioneering policies of Whole Foods and Chipotle, awareness-raising books and documentaries such as Michael Pollan's *The Omnivore's Dilemma* and Robert Kenner's *Food, Inc.*, the carrot-and-stick approach of groups like the HSUS and Mercy for Animals, and the dedication of so many compassionate people who demonstrated that campaigns against the confinement of veal calves and breeding sows resonated with the American public. When voters weighed in on several ballot measures to protect farm animals, it made, in the minds of food retail CEOs, public attitudes even more concrete and inescapable. I'll never forget our advertisements in the 2006 Arizona campaign featuring Maricopa County sheriff Joe Arpaio—the self-proclaimed "toughest sheriff in America"—speaking from a kitchen and saying that while he loved pork chops, "what they do to these pigs in gestation crates is cruel." (He has since opted for serving all the inmates exclusively vegetarian meals at the Maricopa County jail and has become a vegetarian himself.)

Even our adversaries unwittingly contributed to the momentum. A 2007 poll commissioned by the American Farm Bureau Federation and conducted by faculty at Oklahoma State University—both long associated with industrial agriculture and disinclined to exaggerate public support for animal welfare—found that 95 percent of Americans believe farm animals should be well cared for. They also found that nearly 90 percent believe food companies that require their suppliers to treat animals well are "doing the right thing." A 2009 study by Michigan State University (MSU)—another longtime supporter of conventional agriculture—found that 60 percent or more of respondents in every US state with pig production would support a legal ban on gestation crates. "I'm confident that a ballot initiative prohibiting the use of gestation stalls would pass in nearly every state in the Union," study author Glynn Tonsor concluded. And our trading partners were also getting on board with reform efforts. A ban on gestation crates in all twenty-eight European Union member states took effect in 2013; and in 2014, Canada and Australia agreed to phase out the crating system, a further sign to global food companies that confinement-style hog factories were wearing out their welcome everywhere.

By 2015, when Walmart made its announcement, Smithfield Foods had converted 70 percent of its company-owned operations to gestation-crate-free, and was well on its way to its 2017 goal of 100 percent. And in the meantime Smithfield had been purchased by a state-supported company in China, meaning its group housing policy is likely to influence the trajectory of pig production in the biggest of all pork-consuming nations. In 2014, Cargill announced that it had converted all of its breeding operations from crates to group housing. Walmart had also been in discussion with its northwest-Arkansas-based neighbor, Tyson Foods—the biggest meat company in the world, with 14.6 percent of its $37.6 billion in annual sales going through Walmart. Tyson said it knew it had to change and was working with its company-owned operations and its contract growers to that end.

Many of these publicly traded companies—both the meat producers and the food retailers—had been increasingly facing pressure from some of the biggest institutional investors in American business. In 2014, Blackrock and Vanguard Financial started supporting shareholder resolutions to urge companies that produce and buy pork to assess the business risks from continuing to rely on crate confinement. These investors move billions of dollars and try to secure a consistent, positive yield for their customers. We reminded them that pork companies reliant on crate confinement aren't thinking ahead, and are putting themselves in a vulnerable position with their customer base by continuing to treat sows so inhumanely. Even the World Bank has warned companies that they put investors at risk by failing to keep up with consumer expectations of animal welfare. And proxy advisory firms such as ISS and Glass Lewis, which advise institutional investors on how to vote, note that ignoring the problem isn't a sustainable option. "If gestation crates are not part of the lingua franca of most investors," ISS notes, "long-term risk certainly is."

This strategy is having an impact. Institutional investors such as MetLife and Ameriprise Financial helped convince Tyson Foods to make its first-ever announcement that it sees the future of sow housing as not involving gestation crates. And it's investors like these who may ultimately play a crucial role in convincing the laggards—Seaboard Foods and Triumph Foods, for example—to ditch the crating system. My friend Jeremy Coller, with his eponymous investment company pushing around billions in the markets, has been working with the biggest institutional investors throughout the world, such as pension funds and insurance companies, and urging them to adhere to a series of farm animal welfare screens, including no extreme confinement. With that kind of capital suddenly at risk, food companies pay attention.

For those clinging to old, crude ways, there really are no escape routes with an informed citizenry. Consumer sentiment should alone be enough to lever a change in the way the pig industry goes about its business; but when that's reinforced with the demands of food retailers, institutional investors, pollsters, politicians, scientists, and other influential people, you know the end of the era of gestation crate confinement looms. It's almost as if the old guard is in a box, unable to make a turn and see what's best for them. The only humane thing to do is show them a way into the light and fresh air of the burgeoning humane economy.

Assault and Battery Cage

While I was excited to receive Carl Icahn's surprising offer in 2011, I was just as startled to get a friendly call from Marcus Rust in the fall of 2014. Unlike Icahn, Rust was neither a corporate celebrity nor a presence on the Forbes 400 list of the world's wealthiest people. But he was no slouch in business either, as the owner of the second largest egg producer in the nation. Rose Acre Farms had more than twenty-five million birds in hen houses scattered throughout Indiana, Illinois, Iowa, and Missouri, and was therefore one of the biggest corn and feed producers and buyers in the Midwest, too. His birds churned out nearly 6.5 billion eggs a year, and until very recently, just about every member of his flock laid claim to a very tiny piece of real estate within a very small cage.

Rust called to invite me to his headquarters in northern Indiana and tell me of his intention to convert his entire company to cage-free production. It was a bold and game-changing statement, with one of the biggest operators in the field recognizing that cage-confinement operations are on their way out, even in a state like Indiana where he continues to enjoy broad political support. His pledge was even more surprising given our history. Seven years earlier, HSUS had sent an undercover investigator into one of his facilities in Iowa, where our sleuth had obtained footage of hens crammed in small wire battery cages, with some birds dead and mummified alongside the living. It was ugly stuff in an industry where cruelty had become ordinary and routine. That expose had not endeared me or HSUS to Rust, and his deputies did plenty of bare-knuckled counterpunching after we broke the news of our investigation. A year after that tempest, Rust and Rose Acre donated a half million dollars to the campaign to defeat Proposition 2, the landmark California ballot initiative to ban battery cages and extreme confinement systems for breeding sows and veal calves. Rust didn't much like HSUS or our agenda, and he was doing his best to defend his business model.

But when my colleague, Josh Balk, and I flew to Indianapolis and made the drive a couple of hours north to see Rust and a business partner named Mike Sencer, we knew that the egg industry giant had already had a change in heart. In the two or three years preceding our farm visit, Rust and I had talked a number of times, and there had been a gradual thawing in the relationship. Still, though, Rust's pledge to go cage free over time was not something I could have anticipated on this time frame. I had long assumed that the biggest players in the industry would need to be dragged kicking and screaming into the new world order of humane treatment of hens.

Marcus had us hop in his truck and we sped down a country road, kicking up a trail of dust as we headed a couple of miles to his new cage-free hen house. From the outside, the hen house looked like a warehouse, much like any of the other industrial-style egg production facilities Josh and I had seen through the years. Marcus and one of his managers asked us to sheath ourselves in a white zip-up suit and to slip plastic covers over our shoes, so that we didn't carry pathogens into the place. It was an aviary facility—a big rectangular warehouse with a series of long aisles, with every aisle bound on each side by four rows of shelves. The birds used the shelves for perching, sleeping, playing, and nesting. There were no cages to be seen, and the birds could walk, hop, and fly back and forth from the floor to the shelf and from shelf to shelf or just park themselves wherever they wished.

With the surging demand for better living conditions for birds, aviary systems have become common in the egg industry, enabling major producers to house tens of thousands of birds indoors. Rust and his team had been studying the behavior of the birds with this new kind of freedom, and had modified the aviary to make better use of floor space and to deal with the problem of the birds laying eggs where waste had accumulated. There were about twenty thousand birds in the building, and it was a beehive of activity. It wasn't a perfect life, but it was so much better than a life where they were shoulder to shoulder with six or seven other birds in a cage about the size of a microwave. "I never gave much thought to the issue of animal welfare, and I was running a business, just like the other people in egg production," Rust told me, as he pointed over to the birds. "But I can make this work."

It turned out that Rose Acre and Mike Sencer's company, Hidden Villa, were also hatching plans to build cage-free facilities in Arizona and Texas, where the weather is warmer and the buildings do not need to be heated in winter. In Hawaii, they were planning on constructing cage-free operations for a million birds, making Hawaii the first completely cage-free state. As we drove back to the main headquarters from his cage-free barns, Rust pointed to two buildings, and he said he was going to repurpose them. "Those are for Kroger [the nation's second-largest grocer]—all cage-free houses." He said it would take time, and be an enormous expense, but he was prepared to convert every one of his facilities to cage free. It was a thrill to see a long-time adversary turn into an ally, and to recognize that a leader in the old, inhumane economy was ready to find a new path forward too.

And Rust wasn't the only one. Not long after visiting Marcus Rust, I received a message that Dave Rettig, the CEO of Rembrandt—the third largest egg producer in the country—wanted to come see me. Rettig and his lawyer, along with two officials from the United Egg Producers (UEP), the trade association that represents the vast majority of large producers, visited me and Josh Balk at our downtown office in Washington, D.C. It was also a bit of a head-spinning moment, since, as with Rose Acre, we had conducted an undercover investigation of Rembrandt's operations too. There are good outcomes from the sunlight of these investigations, but a warm feeling between the investigators and the investigated is not often one of them. At some level, it was remarkable that these CEOs were even talking with us.

Rembrandt mostly provides liquid eggs to many food manufacturers, and these "breaker operations," as they are known because they break eggs into liquid form, are well known for often not even meeting the UEP's minimal animal welfare standards. Rembrandt had been a notorious name in our quarters because it kept millions of its hens in forty-eight-square-inch allotments. It was hard to imagine anything worse than the conventional cages that offer each bird just sixty-seven square inches. But shrink them further and jam in more birds. That's the can-of-sardines effect birds get with Rembrandt and the other breaker operations, except that unlike the sardines, the animals packed so closely together are still alive. As with Marcus Rust, however, Rettig told us that he too was ready to convert his company and make it cage free. But he needed help. He needed customers, and that's where he thought we could help, given our connection to consumers and the influence we have with so many major food retailers. We were only too happy to oblige.

In these constructive, friendly encounters—sitting next to Marcus Rust in his truck and across the table from Dave Rettig in Washington—it became clear that the smart minds and money in the egg industry had figured out there was no future in cages. According to their way of thinking, they had to adapt to that reality and the sooner the better. They wanted to get ahead of the rising consumer tide of concern about animal well-being and food safety. And perhaps they also felt a sense of responsibility to do things better, since the conventional practices in their industry exacted such a mighty toll on these innocent creatures. I don't think many people of conscience, once they become alert to the suffering of animals, want to be a part of it. Yet, there's something amazing to behold when two of the three top players in an industry driven for years by considerations so focused on dollars and cents want to make such a dramatic pivot. Egg production had become, in just a few short years, one of the flashpoints in the national debate over the proper treatment of animals, and with birds moving from cages to cage-free environments, we were starting to see change.

I've visited many farms with free-range hens, such as Eliza MacLean's Cane Creek Farm in central North Carolina, and there's something joyful about seeing the hens at work and play. The hens forage, flock, roost, and chase one another in grass that's chest high. At Cane Creek, and on so many other farms that give the birds some measure of freedom, the animals have lives worth living. For Eliza and other farmers who practice this kind of mixed animal husbandry, the birds provide a livelihood; for society, they produce an inexpensive, sustainable, protein-rich food.

In November 2012, I visited the rural regions of Andhra Pradesh, a state in central India, with my colleagues from HSI for a project we support that provides laying hens to rural communities. Many of the people we met there live on a dollar a day. By acquiring laying hens who roam and forage and need no supplemental feed, these people, scraping by each day for food and the other bare necessities of life, have something that can change their fortunes. The villagers collect the eggs, eating some and selling the others at local markets for cash. It's a win-win—the chickens do just fine, and people living on a knife's edge are better able to get by.

But it's far from a winning situation for hens in most other places where they're conscripted for food production. After visiting the rural villages with the free-range hens, our India Director N.G. Jayasimha and I walked right onto three battery cage operations in Andhra Pradesh. The cage facilities were appalling, with hens crammed into spaces so small they could barely move, the air thick with flies and ammonia, and the ceiling covered with cobwebs—thick and heavy with fecal dust—that hung down like drapes. Within a year of that visit, we successfully petitioned the Animal Welfare Board of India to declare confinement cages illegal under the country's Prevention of Cruelty Act. But enforcement has been spotty, and we're still working to shutter the nation's remaining battery cage facilities.

In the United States, upwards of 85 percent of laying hens live in battery cage confinement systems, and until a few years ago, there were no laws at all even stipulating that the birds should be able to move. It's a highly consolidated industry, with just one hundred or so companies, including Rose Acre and Rembrandt, accountable for upwards of 85 percent of the 90 billion eggs produced in the country.

I had my first close-up view of an industrial-style egg operation in the United States in early 2008. The visit did not come at the invitation of the owner; instead, it came from a fed-up neighbor, who could no longer stand the stench and the swarm of flies originating from the factory farm next door. In the run-up to the 2008 vote on California's Proposition 2—the statewide anti-confinement ballot measure for farm animals—Dave Long gave me an unauthorized tour of the Riverside County egg production facility that had turned his life upside down. We couldn't gain access to the two modern-style confinement houses—each of which had more than 100,000 birds in cages stacked five high. But taking a ride on Dave's John Deere ATV, we were able to get a close look at the 400,000 or so hens divided up among about twenty old chicken houses. In each "house"—consisting of just a roof and open sides—there were five columns of cages running the length of each long barn. The columns were on stilts and comprised two cages joined at the sides, each angled down and out so the eggs would roll and collect in gutter-type metal pans. The two side-by-side cages were stacked two high, and there were two to four birds in each cage. The birds in the lower-level cages were much darker than the birds above, not because they were a different breed, but because manure fell through the wire cages above and pelted them every day, giving them a dingy hue. For them, there was no escape from the waste—it was coming at them from above and below, with a growing three-foot mound of feces just inches from the floor of the cages. With clouds of flies buzzing, it was hard for me to endure it for even fifteen minutes. How must it feel for these birds, who know nothing else of the world but this miserable place?

Prop 2 was a sort of political earthquake for both animal welfare and agriculture. That vote, according to Dave Long, led his neighbor to modify the archaic outdoor caging system once the law took effect in 2015, bringing the birds, and Dave, some relief. But the ballot initiative's tremors were felt far beyond California's boundaries, especially after lawmakers passed a follow-up measure requiring out-of-state producers selling eggs in the state to maintain their hens in housing systems compliant with Prop 2. Iowa, in particular, felt the shock, given that it's the top egg-producing state with sixty million hens. With just three million people there, fifty-seven million of those hens are working for people outside the state, including California consumers.

A couple of years after voters passed Prop 2, but before it took effect (it had a six-year phase-in effort to allow the farmers to convert their systems), I was able to visit Iowa's biggest egg farm—a ten million hen facility—on a trip organized by a man who became an unlikely friend and unexpected problem solver. Jerry Crawford, an influential Iowa attorney with a number of major egg industry clients, got me authorized access to Wright County Egg (formerly DeCoster Egg Farms), then one of the nation's most notorious factory farms. Even for the well-connected Crawford, who has a reputation for being able to make almost anything happen in Iowa, that was a feat. He had been prominent in the Iowa presidential caucus campaigns for Bill Clinton, Al Gore, John Kerry, and Hillary Clinton. I got to know him through the HSUS board chairman and attorney Rick Bernthal. When Bernthal first mentioned my name to Crawford, the Iowa native's face went flush. He had heard much of the claptrap about the HSUS being anti-agriculture and an enemy of farmers. After some back and forth, though, Bernthal convinced Crawford to at least sit down with me. Despite our very different worldviews and our contrasting experiences with the egg industry, we hit it off. After meeting a few more times, Crawford was insistent that I see egg production firsthand and meet the people running the egg farms—thinking perhaps that a closer, more personal inspection might soften me up.

I flew to his hometown of Des Moines, and we headed north for nearly ninety minutes to Wright County Egg, passing through an unending patchwork of corn and soybean fields, interrupted by the occasional farm house or towering gray windmill. We met Peter DeCoster, who was about forty years old with a dark head of hair, some gray-and-white stubble on his face, and a wiry frame, at the company's small brick office building. Peter was a key operator, though his father Jack was the owner. Jack DeCoster had been in trouble with the law for years for polluting the air and water and violating his workers' rights—this in a state known for lax oversight of agribusiness—and he was considered a pariah even within his industry.

Crawford and I had our site visit, as it happened, just months after a summer 2010 recall of 500 million eggs from this facility and a nearby, separately owned operation called Hillandale Farms. The recall action was triggered by salmonella poisoning traced back to eggs at these operations. It was the largest egg recall ever, and it had roiled not only the DeCoster clan and business, but also the entire egg industry. Nationwide, the tainted eggs from the Wright County and Hillandale operations had sickened what some reports estimated as more than fifty-six thousand people, with ten deaths. That controversy and national food safety scare provided another reason for California lawmakers to stipulate that any eggs sold in their state had to be compliant with Prop 2. Iowa eggs were very much on their minds.

Peter DeCoster told us the hens were divided between ninety buildings, scattered over hundreds of acres, and he asked that we hop in our car and follow him to one of the houses. After a five-minute ride, we pulled up to an enormous, nondescript building, the length of a football field, and he gave us instructions to slip on plastic booties and protective suits. We looked like we were all on the same team—head to toe in white, matching the color of the birds in the buildings. The younger DeCoster was as pleasant as could be, though understandably anxious about my being there. The company had only recently resumed sales after the recall, which had drawn much unwanted attention and landed both father and son before a congressional committee. The next step for the DeCosters was a federal criminal case against them and their co-defendants; later, both DeCosters were sentenced to three months in jail and fined $7 million. With all that coming his way, you have to give Peter DeCoster credit for letting me have an inside view—hospitality I can't imagine his father would have ever extended.

We weren't inside long before DeCoster scooped up a hen who

had escaped a cage. He lowered himself down on one knee, holding the hen in his left arm. As the bird watched, he dragged his finger along the ground in front of her in a back-and-forth manner, doing this for about twenty seconds before letting go of the hen. She splayed out, motionless and looking straight ahead with a vacant stare. He had hypnotized the bird. It was a harmless mind game, and probably a much better place to be—physically and mentally—than in the cages with all the other birds. After a couple minutes, as Crawford and I watched in amazement, DeCoster snapped his fingers and she came out of it, trotting away and again enjoying what would almost certainly be her last dash of freedom before she was locked up. Crawford told me that Jack DeCoster could outdo even that—walking into a henhouse filled with the noise of 100,000 birds and make a deep guttural sound that would silence every last bird. Whatever else I was about to see, it was plain that the DeCosters knew a thing or two about chickens.

Peter DeCoster walked us down the rows, offering little observations above the clatter of the birds, as stacks of cages towered above us to the left and right. The birds never leave their cages and are denied any opportunity to live like birds—to forage, to feel grass beneath their feet, or even to see the sky or the sun. As I looked around and surveyed this mass of animal life, white feathers everywhere, it occurred to me that you'd have to be in a trance of your own to think this is acceptable. If only a snap of my fingers could have taken them out of all of this.

As if the twelve to eighteen months of confinement aren't bad enough, the birds then get shipped off to slaughter. In 2015, the HSUS released the first-ever investigation of a spent-hen slaughter plant—one that kills laying hens whose bodies are too worn out to keep producing enough eggs. One of our brave investigators went undercover at Butterfield Foods Co., a Minnesota slaughter plant in Watonwan County. That person documented hens shipped from battery cage operations throughout the United States and Canada, many arriving severely weakened or debilitated by broken bones and starvation. The industry has a knack for coining twisted mottos; in this case, it was "We Love Old Hens."

Our investigator's report documented how the plant receives hens who had for hours on end been locked inside transport cages even smaller than the battery cages, and how the birds were crushed together more tightly than in the barns. If hens in battery cages are like eight people jammed in a tiny elevator 24/7, add four or five more people and shrink the elevator, and you get a sense of what it's like for those birds in transport. They are trucked to slaughter through all kinds of weather conditions; no food or water is provided. Our investigator found that hens delivered to the plant over the weekend are simply left to languish on the trucks until the killing shifts resume again on Monday. Exposed to the Minnesota winter, those in the outer rows froze to death.

When it came time for the slaughter, our investigator saw workers removing hens from their crates by jabbing metal hooks into the densely packed transport cages to rip them out of the cages by their legs. The workers then hung these hens—still fully conscious—upside down on the slaughter line. After being shackled, the row of upside-down hens moved through an electrified trough of water designed to stun them, although that was not always the outcome. Every day, hens were scalded alive as they were forced upside down into tanks of scorching hot water. In just one thirty-minute period, our investigator witnessed approximately forty-five birds drowned or scalded alive. The Butterfield Foods plant—like nearly all others in the poultry industry—kills hens in an archaic process that would be illegal under federal law if the animals were cattle or pigs. But since the USDA exempts birds from the federal Humane Methods of Slaughter Act—literally removing them from the law's definition of "animal"—there is not even a requirement that these laying hens be rendered insensible to pain before they're killed. The plant paid two to three cents per bird—providing a small profit for the egg producers who might otherwise have sent the birds to the landfill. With almost no value to the people involved in this enterprise, and in vast numbers, the hens were viewed as little more than trash.

The misery of intensive confinement back at the hen houses is compounded when you consider that the birds cannot establish their pecking order in this scrum, and must draw their breath from an ammonia-filled atmosphere of their own collective making. The typical factory farmers keep the lights on for sixteen or seventeen hours a day during the laying cycle. If there were a contest to develop the most wicked and cold-hearted means of egg production, this just might be it.

After our tour of the DeCosters' farm, I wasn't sure if Jerry Crawford was disturbed by what he'd just seen—after all, this was sort of a routine experience for a guy who'd been an industry insider for years. But the whole scene sure troubled me, and I came away convinced that this entire disgraceful system of conventional egg production had to change. Sure, the public was on to Jack DeCoster already, and many of his peers considered him a rogue operator. But there was something fundamentally wrong with the standard-issue cages even for producers playing by the rules. These egg producers had all justified their extreme treatment of animals with a series of rationalizations: they protect the birds by keeping predators away; they provide them with adequate food and water; and birds live in flocks and instinctively crowd together. And since, after all, the hens are mere animals, the producers were doing it all in the name of humanity—providing hungry consumers with an inexpensive staple food. Crawford himself had offered that latter argument to me many times, as we sparred about what should be best practices in the egg industry.

Not long after our tour in Wright County, Crawford and I traveled to Europe for a comparative look. We visited several egg farms in the United Kingdom and the Netherlands, all in the process of transitioning from battery cages to colony cages as a result of reforms by the European Union. Each colony cage housed about sixty birds apiece, but provided about double the space per bird as conventional cages, along with nests, perches, and scratch pads. By Crawford's arrangement, our guides came from Big Dutchman, one of the two big global manufacturers of laying hen housing systems—whether battery cages, colony cages, or aviary-type cage-free systems.

One of our first farm stops was an operation run by two brothers, who owned about 450,000 birds divided between four houses. The aptly named Byrd brothers had two conventional battery cage operations where the manure stacks up at the bottom of the barn for the twelve- to eighteen-month lifespan of the birds. Only when they remove the birds do they clean the barn. They also had a third barn in which the birds lived in conventional cages, but with wide plastic belts below the cages to carry the manure out of the building. A fourth barn has colony cages, with the additional space and enrichments, plus the belts to remove manure.

The Byrd brothers told us that they had fought against the European Union regulation to ban battery cages with all they had, believing that their systems were adequate for the birds and delivered eggs to the table at a reasonable price. But they were law-abiding men and reluctantly agreed to start retrofitting their hen houses, one by one, with the systems built by Big Dutchman, to comply with the new rules. After converting one barn, and seeing it in operation, the Byrds became converts themselves. Mortality among the birds was much lower, and so was the cost of feed. The in-barn workers didn't need any convincing either, because the air quality was considerably better. And so obvious were the improved animal welfare conditions that everybody, including the Byrd brothers, suddenly felt better about the business they were in. Who wants to spend eight or more hours a day in the company of thousands of suffering, immobilized creatures?

In seeing all of this, especially the enthusiasm of the Byrd brothers, Jerry Crawford began to come around as well, realizing that battery cages were inferior to the larger colony model. The impulse to do better and be a better person and to be on the side of innovation is a strong one, as I'd see time and again in my work in agriculture and other sectors of the economy. Crawford later arranged a secret meeting for me with Chad Gregory, who was soon to take over from his father as CEO of the UEP, the egg industry association. As in the case of my first visit with Crawford, the initial meeting with Gregory had an awkward feel, since he and I had been bitter adversaries over Prop 2 and in other fights over the treatment of hens. But in the comfortable setting of Crawford's second home in Arizona, we relaxed, and it didn't take long before we were having a serious discussion about the future of the egg industry. We met a few more times, and with the Prop 2 result still fresh in his mind, Gregory said he wanted to lead the industry to change. He, too, had seen the colony cages in Europe, and he could see that they were far superior to the systems in the United States. But making the switch was the challenge—both the costs of retrofitting thousands of confinement barns, and the task of convincing egg producers that all the disruption and change was worth it. He wanted to lead, and he was ready for the task.

Gregory did a whirlwind tour of the US egg industry, visiting producers in all the major producing states, and with prodding and cajoling, the vast majority of them got on board with the plan to move away from battery cages. The HSUS and UEP formally agreed to join together in an effort to end the era of battery cages.

It was through that unlikely association that I had originally opened up some real dialogue with Marcus Rust and a number of other people in the industry, starting in 2011. During that process, despite my antipathy for the systems they oversaw, I learned that these egg producers are good, hard-working people, who chose their production methods to boost efficiency and not to purposefully make life miserable for their hens. From their perspective as farmers running a business, battery cages have been a smashing success. They can make good money, and in some cases become extremely wealthy, by producing a product at a high volume and a low cost—$2 or less for a dozen eggs had been the going rate for a number of years. Without knowing much about how the birds were raised, consumers have rewarded them with brisk demand, with the average American in 2014 consuming 264 eggs—a number that sounds almost impossibly large until you realize that eggs are baked or blended into muffins, cakes, pastas, mayonnaise, cookies, and all sorts of other products. Each hen produces about that same number of eggs—about 260 per year. So, if you eat eggs and you have a conventional diet, you keep a chicken busy every day of the year. But, unless you specifically buy cage-free eggs, she's almost certainly crammed into a small cage every minute of her misery-ridden life.

Chad Gregory and I made a much-hyped and widely lauded announcement together at a press conference on Capitol Hill, and in the months that followed, spent hundreds of hours lobbying lawmakers to adopt the terms of the agreement. Producers like Marcus Rust called on their lawmakers to join our fight. It was the unlikeliest of circumstances for us, as long-time nemeses walked the halls of Congress and advocated for legislation calling for a transition to the new housing systems—setting a minimum standard to double the space per bird, provide enrichment, and require on-carton disclosure of production methods. What the UEP didn't want was a voluntary program that would result in a conscientious producer switching to a more humane system only to be undercut by a competitor who kept the battery cages in place. The egg producers wanted one set of rules that would be fair for everybody. That made sense to me, and that was one reason why we swung behind the federal reform effort.

In the end, despite the underlying wisdom of the reform and the unprecedented union of the HSUS and the egg industry, Congress failed to act. That was largely due to the meddling influence of the pork and beef industries, which were hostile to setting a precedent for any federal regulation of farm animal welfare. The legislation was never defeated—it just was denied an up-or-down vote after amassing nearly unanimous Democratic support and considerable backing among Republicans. It was a bitter reminder that Congress was often unable or unwilling to act even when handed an agreement tailor-made to satisfy the key stakeholders.

All was not lost, however, since the idea of sensible reform is hard to suppress and often takes flight through other pathways. California's new hen-welfare laws were set to take effect in 2015 and would apply not only to egg producers but also to egg retailers, and thus create markets for egg farmers committed to more humane systems. Major retailers came on board as a matter of law and corporate social responsibility. In December 2014, on the eve of the implementation of Prop 2, Starbucks—with fifteen thousand outlets—agreed to source all of its eggs from cage-free farms by 2020. I could see the momentum building for the idea of skipping the shift from battery cages to colony cages, and for the industry to make the leap entirely to cage free.

In early 2015, the nation's biggest food service operators—Compass Group, Sodexo, and Aramark—also agreed to go cage free; they buy more than a billion eggs a year, guaranteeing that 3.5 million birds would get out of cages. Dunkin' Donuts also weighed in and signaled a change. It was just a few months later when Walmart made its "Five Freedoms" announcement. With gestation crates and veal crates on the way out, battery cage confinement of hens would soon become the last bastion of intensive caging of animals, and who wanted to be that kind of outlier? While Walmart didn't announce a timeline for its shift to cage free, it was clear that's where the company was moving.

But as with the gestation crate and the pork industry, the biggest announcement affecting the future of the egg industry came from, again, McDonald's. In September 2015, just four months after the Walmart declaration, the company announced it would go entirely cage free over the next decade for all of its restaurants in the United

States and Canada (it had already opted for cage free in Europe some years earlier). That announcement had even more force because McDonald's had been at the center of a consortium of companies, research scientists, and the egg industry that backed a multiyear, multimillion dollar comparative study of laying hen housing systems. By every indication, the study seemed preordained to conclude that colony cages, rather than cage-free production, best balanced economic and humane considerations. We knew McDonald's was going to do something to improve the lives of hens, but we were concerned that the company might move too slowly and prolong the fight over colony cages versus cage free.

For years, my colleague Paul Shapiro had trekked to every McDonald's shareholder meeting and politely but persistently made the case that the era of intensive confinement had to end, and that cage-free systems would ultimately be the only housing systems acceptable to the public. In between the shareholder meetings, Shapiro had been lobbying and negotiating with McDonald's, yet growing increasingly concerned about the direction of the industry study. So it was with some measure of alarm that a vice president of the company told him she and others in the company had a final decision to share. Shapiro told me he had "a lump in his throat" as he dialed in. The lead spokesperson with McDonald's dispensed with any formalities once the call commenced. "We don't want to bury the lede," she told him. "We're announcing next week that we're going 100 percent cage free within ten years."

It was a long time frame, but it was the right end point. Shapiro said he was so happy that he hardly knew what to say, since he and several other members of our team had devoted their adult lives to ending battery cage confinement, and all of us who'd been in this battle felt that McDonald's pledge would seal the trajectory of this debate. From this point forward, we'd be negotiating terms and time frames, not debating the fundamentals any longer.

With the company buying and then selling two billion eggs a year—about 3 percent of all eggs in the United States—its impact would reverberate throughout the producer and food retail sectors of the economy. As with pork purchases, when the biggest, most cost-conscious companies say they can make the switch and that they won't have to raise prices for consumers, it leaves no wiggle room for other companies. They too would announce similar policies, as we saw after McDonald's led the sector in saying that gestation crates would no longer be acceptable. McDonald's made the announcement just three weeks before it rolled out an all-day breakfast menu—the segment of its offerings that relied most heavily on eggs. Two months later, Taco Bell—part of the Yum! Brands food empire that was an outlier for years and had made no animal-welfare pledges—announced it, too, would go 100 percent cage free. It would set a new standard by saying it would get there in a single year—quite a feat for a company with six thousand outlets and an array of breakfast offerings that featured eggs. By the end of 2015, Craig Jelinek, CEO of Costco—which sells more eggs than McDonald's—told me the company would be closing in on 100 percent cage free sales by the end of 2018.

McDonald's, Costco, Taco Bell, Sodexo, and the major food service and retail companies didn't make their decision on a lark, or simply because the HSUS demanded it. If food retailers were to shift their procurement strategies, they'd need egg producers to supply them. On the heels of the McDonald's announcement, Glenn Hickman, owner of a top-ten egg producer in Arizona, also said his company was going to make the transition over time to entirely cage-free production. "Cage free is just the next logical step," Hickman said. "It's the future of the industry." And Dave Rettig's declaration, that "cage-free egg production will be the company's standard" came just a month after the McDonald's announcement. "We welcome the growing movement of major food companies switching exclusively to cage-free eggs," said the guy who oversaw an operation that had been ruthlessly tough on the birds: "With a reasonable timeline, we can meet any demand, and we're eager to propel our clients into the cage-free future." Steve Herbruck, one of Michigan's biggest producers and also a convert to a cage-free future, had moved four million of his seven million birds to cage-free houses by the time McDonald's made its announcement. As more companies got on board, he'd be in the best position to accommodate them, having methodically built his cage-free flock because he could see the way the debate was going. For him and so many others, they'd never build another cage facility. All the new production would be cage free.

Chad Gregory—who led a $10 million fight against Prop 2 in 2008—told me that even he came to accept that the future of the industry was cage free. In a seven-year period—from the landslide vote in California in favor of Proposition 2 to the present—the thinking of the food retail sector and the egg industry had gone from all-out combat with us to open dialogue, and then ultimately to broad acceptance of a cage-free future.

Every business has to earn its customers today and tomorrow, and that goes for producers, trade associations, and retailers in the food business. In the marketplace, there are few entitlements. Even the biggest, most established brands are vulnerable to competition from upstarts and the impact of evolving values. Cruelty baked or boiled into the business model is, more than ever, a risk not worth assuming. Walmart's own customer surveys showed that 77 percent of its customers will "increase their trust" and 66 percent will "increase their likelihood to shop from a retailer that ensures humane treatment of livestock." And when public opinion breaks against a long-established practice, it can do so very quickly, as we've seen in the pork, egg, and veal industries. More people than ever are inclined to take their business elsewhere if they find out that animals are harmed in making their favorite products. And as more companies get on board and adopt animal welfare reforms, the resisters also become pariahs. As many companies inside and outside of the animal-protection debate had learned, it doesn't take a full-fledged consumer revolt or exodus to disrupt a business either. Even 10 or 15 percent of aggrieved or angry consumers can shrink sales, agitate via social media, and conduct protests, with the potential to irreparably damage a brand. Every CEO must be alert when powerful ideas migrate from the margins to the mainstream, and they must be ready to adapt or be left behind.

Remember when the auto makers fought seat belts as a disaster for their industry? Now safety is a key selling point in the car business. The very idea of incorporating animal welfare as a core value of industrial agriculture was viewed as naïve and impractical just a few short years ago, and even a bit of heresy for anyone in production or food retail to adopt. Now it's a selling point if a company has adopted practices that alleviate harm to animals. In a culture where there is widespread antipathy to cruelty, that thinking must be infused into the business plan. CEOs are turning social concern into economic opportunity: Chipotle and Whole Foods did so as pioneers; Walmart and McDonald's did so to align their business model with the evolving values of the American public; and Marcus Rust, Dave Rettig, and other producers did so because they didn't want to be left with billions of eggs and nowhere to sell them. Doing the right thing is no longer viewed as sacrificing efficiency or profit, but as good business. It's a brand builder and a share builder and a hedge against future risk. And by embracing that value system, no CEO ever has to fret about getting a call out of the blue from a guy like Carl Icahn or a request to fight a ballot initiative or a federal bill or regulation to upend his or her business. And no CEO wants consumers to view the company as a laggard and a contributor to cruelty. When executives lead and do good things for society, they earn, and that's precisely because enlightened consumers demand it.

CHAPTER THREE

The Chicken or the Egg—or Neither?

"We shall escape the absurdity of growing a whole chicken in order to eat the breast or wing, by growing these parts separately under a suitable medium. Synthetic food will . . . from the outset be practically indistinguishable from natural products."

—SIR WINSTON CHURCHILL, 1931, in
an essay titled "50 Years Hence"

Meat With, and Without, a Heart

Winston Churchill missed the mark by a few decades, but the rest of his vision was characteristically bold and accurate. For those consumers not satisfied with more humane methods of raising and slaughtering animals—as significant as they've been in the last five years—there are new food production strategies not just on the drawing board, but on the milk and cheese shelves and even in the meat case. The meat of the future has arrived, at long last removing the whiff of suffering or the stain of blood from the food we eat, and it promises to be one of modern science's finest hours.

Andras Forgacs came to the downtown office of the HSUS in Washington, D.C., to tempt me into eating meat for the first time in thirty years. But Forgacs isn't a rancher or a meat industry rep, and the meat didn't arrive sealed tight in shrink-wrap. Rather, Forgacs is a Harvard-educated scientist, and his meats came in the form of bite-sized appetizers, carried in his shoulder bag. He and a few biofabrication scientists with his company had found a way to take a pinprick's worth of muscle cells from a steer or boar and turn it into a small mound of edible flesh: "Lab-grown meat. No animals harmed. Made in Brooklyn."

Forgacs' company, Modern Meadow, is a pioneer in the experimental world of in vitro meat and leather production. The process is not nearly as difficult as salvaging and replicating 10,000-year-old, fossilized DNA from a wooly mammoth or mastodon. Forgacs isn't interested in making Jurassic pork. Quite the opposite, he's fabricating meat from some of the most abundant species on the planet.

Forgacs is part of a new universe of inventors who are out to change our lives, and the agricultural economy, in the most fundamental of ways. A group of scientists founded Memphis Meats to grow meat in steel tanks. Three Indian-Americans have started a Silicon Valley–based company called Muufri to create milk and cheese from genetically engineered yeast. A few months before I met Forgacs, Dr. Mark Post and his team at Maastricht University in the Netherlands cooked up the world's first lab-grown hamburger and served it in London to a group of journalists. For the diners, it was a meal to remember, though not because the futuristic fare was perfectly to their liking. Even though it will need years of refining, cultured meat is one of those bellwether products, like the first PC or those early-model cell phones of the 1980s, whose mere existence portends that bigger changes are on the way. Dr. Post's burger-fabrication operation originally had a single main backer, but an important one with quite a track record as an innovator: Google co-founder and billionaire Sergey Brin, who said he invested because "when you see how these cows are treated, it's certainly something I'm not comfortable with." He doesn't hold the view that developing cultured meat is impossible or naïve. "If what you're doing is not seen by some people as science fiction," Brin told the press, "it's probably not transformative enough."

Modern Meadow's Forgacs had good things to say about Dr. Post and his work, but stressed that the Dutch operation is more experimental, using a different fabrication process: "He is recreating muscle fibers, one fiber at a time," and then binding them together. Forgacs thinks his own method of tissue engineering lends itself more readily to commercial application. If that sounds like a classic race between inventors, it is. Dr. Post now says he can get the wholesale cost down to $11 for a burger.

Forgacs says his burger, over time, will become competitive on price, too. He and his team start the fabrication process by taking a muscle biopsy. "Once you isolate the cells, you grow them in a cell culture," he tells me. "It is a soup that contains the vitamins, minerals, sugar, and all of the other nutrients needed for growth." It gains mass through cell division—the same biological process that produces growth in an animal.

But Forgacs' process needs fewer inputs in comparison to conventional meat production because, with no need for legs, hooves, or hair, and no pumping heart or lungs, the caloric burn rate is dramatically lower. It's a precise rifle shot, with all growth directed to edible meat and usable leather.

Listening to Forgacs, my mind went back to the observation decades ago by writer Frances Moore Lappé in *Diet for a Small Planet* that "animals are protein factories in reverse." It can take twelve pounds of grain to put one pound of flesh on a cow, to say nothing of the incredible water, land, and fossil fuel requirements at every stage. Forgacs must still use plant material to grow animal flesh, but he does it all so much more efficiently. And even that hugely understates the positives.

Modern Meadow's operation also generates negligible waste—just some dead cells and other biological residue. By contrast, the Environmental Protection Agency reports that the nation's 18,800 Confined Animal Feeding Operations (CAFOs)—facilities with large numbers of cattle, pigs, and chickens confined inside—generate five hundred million tons of manure annually. That's about one trillion pounds, or three times the total weight of meat, milk, and eggs Americans consume every year. And animal agriculture produces another 243 million metric tons of carbon dioxide equivalent in greenhouse gas emissions—making it one of the largest emitters on the planet. A European Union study predicts lab-grown meat could reduce land use by 99.7 percent, drain 94 percent fewer gallons of water from aquifers and rivers, and reduce greenhouse gas emissions by 98.8 percent.

It's fitting, too, that all of this unfolds in Brooklyn, once the site of countless stockyards, packing plants, and meat markets. In nineteenth century America, our cities were crowded with animals, including livestock. Raising farm animals not far from consumers, or transporting them alive and slaughtering them close to home, was a practical necessity, because in transit "beef and pork went through a series of mutations that rendered them first unpalatable, then inedible, and then dangerously toxic," according to University of Wisconsin historian William Cronon. For more than half of the Industrial Age, and for all time prior to that, most food was "local" and farm-to-table was the standard. After all, our nation was still largely agrarian. Many farmers did their planting and rearing at the edge of the cities or, in some cases, right within them.

But with the nation's westward expansion, settlers could lay claim to fertile land on both sides of the Mississippi River flood plain and across the Great Plains region, the vast mid-section of the continent, stretching from Texas up through the Dakotas and into the prairie provinces of Canada. In the years after the Civil War, sport and market hunters liquidated the bison, and federal troops and disease continued to decimate the Indian tribes. European-Americans didn't mix well with native people or wildlife. With the lands' original inhabitants either eradicated, or pushed into small pockets or reservations, ranchers and farmers could put longhorns and other cattle, sheep, and pigs onto the "open range"—taking advantage of the natural grazing lands that had once supported millions of bison. Or in many areas they could put corn in the ground, supplanting the native prairie grasses, and feed that crop to farm animals.

By the latter part of the nineteenth century, refrigerated rail cars connected these farmers and ranchers to cities swelling with Americans and European immigrants hungering for meat. Increasingly, agriculture was outsourced to these rural regions, as cities turned from farming to manufacturing, construction, finance, and other enterprises associated with an industrial economy.

But the cities still did much of the trading and processing of animals, even if it took cowboys and train conductors to drive the animals there, often over hundreds of miles of terrain. The destination typically was a stockyard, where the different players in the meat industry aggregated the animals and then either traded and shipped them to the East or slaughtered them on-site. Chicago's Union Stockyards, in the city that became known as the "hog butcher to the world," employed twenty-five thousand people at the turn of the twentieth century, and processed more than 80 percent of the meat sold in America. In the 1870s, it boasted "2300 pens on a hundred acres, capable of handling 21,000 head of cattle, 75,000 hogs, 22,000 sheep, and 200 horses, all at the same time." Cincinnati, which specialized in collecting, trading, and slaughtering pigs, was known as "Porkopolis." In Manhattan's Meat Packing District, on the edge of the Hudson River, 250 slaughterhouses turned live animals into whole carcasses and cuts of meat.

Today, urban stockyards have mostly vanished, even in Chicago, Kansas City, Minneapolis, and the other cities of the Midwest. The interstate highway system and the rise of trucking allowed farms to move live animals directly to slaughter plants without need of intermediate trading centers. Over time, slaughter plants migrated to rural regions too, enabled by cheap land, proximity to the farms, and refrigeration. It became cheaper to ship frozen meat to cities than to move live animals.

Today the slaughter plants are gone from Manhattan's Meat Packing District, an area now known for fashion, bars, and restaurants. The same is true in Brooklyn, where young professionals live in converted buildings and lofts alongside start-ups like Modern Meadow. Thanks to Forgacs' company, Brooklyn may be the only geographic jurisdiction to produce more beef in labs than in fields or feedlots.

What Andras Forgacs carried in his shoulder bag was a rounded steak chip—resembling the shape and circumference of the petri dish it had grown in. "It's very low in fat and high in protein, very healthy and shelf stable, very food safe," he told me, describing it as a kind of beef jerky. "Why grow animal products from sentient animals when you can grow them from cells?"

"The science of cell culture has been around for 100-plus years, but the technology has accelerated over the last several decades," Forgacs explained. "We now know how to grow cells in a lab very easily. In fact, there are many products predicated on cell culture already, like yogurt, which is a food product that involves culturing lacto bacillus. Yeast is a cell culture product. A whole bunch of drugs in the biotech industry are made of cells that grow in vats, like insulin for medicine."

Modern Meadow is Forgacs's second biotechnology start-up. He founded a company called Organovo, which does 3-D bioprinting of medical tissues—using a machine to layer cells and replicate livers, kidneys, skin tissue, and other living organs. By mimicking actual human biology systems and key parts, scientists can more effectively test drugs—better than testing on animals, whose physiology often doesn't mimic the effects of drugs or toxins in a human system. The other purpose of the printing is to duplicate organs for use in transplants, or to fix organs, like an infarcted heart. "If we can make medical-grade human tissues that can solve important issues in human medicine and drug research," he said, "then why can't we also create animal tissues that can help with more global environmental and animal issues." That was the logic that led the scientist into the one trillion dollar meat business.

In the era of the Internet, we hardly need reminding how often technologies can upend long-standing industries, ultimately for the good of all. "We are entrepreneurs," Forgacs told me. "We are very motivated by solving global problems, but also quite aware that there are big opportunities and big potential markets. We are creating a new industry."

Today, the world raises seventy-seven billion land animals a year for food—eleven cows, pigs, chickens, and other creatures for every person. Per capita consumption of animal products is declining slightly in the United States and Europe—though not nearly enough, as Forgacs points out, to offset increases elsewhere. His wake-up moment came in Shanghai, where the rate of meat consumption was going up and anything of quality was flown in from Latin America or Australia. "In Shanghai, why were people eating meat that was transported halfway around the world?" With wages increasing in China, and a growing middle class eating more meat, the resource-related problems would become even more acute over time. "You cannot double these [meat consumption] numbers," Forgacs warned. "We are already stretching the carrying capacity of the planet."

Forgacs fished from his bag a few barbecue-flavored meat chips, each carefully packaged. He placed some on a plate in front of me. Then he pulled out his iPhone and showed me a video of Sergey Brin trying one, followed by former US Agriculture Secretary Ann Veneman—an act that might have drawn indignation from the cattlemen's association if she'd still been in office. Both of these tasters seemed to think the meat palatable, but also struggled for the words to describe it. Neither of them jumped out of their chair demanding "a whole bag of these things!" But that wasn't the point or the expectation. These tastings were not about dazzling anyone on palatability, but about crossing the threshold from "this is ridiculous" to "I can see this happening." So many new ideas seem all but impossible. Then inventors work out the kinks and the first products are made ready for commercial use. Then it's in the hands of the early adopters, who embrace new ideas and technologies and, if things work well, prepare the way for their broader acceptance.

I asked Forgacs about the challenges he'd face in making his product commercially successful: public acceptance of eating a lab-grown product, palatability and taste, production on a viable scale, and overcoming political and other interference from the meat industry. Food is our first necessity, and there have been a series of revolutions in the annals of eating. Domesticating wild animals and using them for food and agriculture-related labor was one of the most transformative processes in all of human history. The discovery of micronutrients was a more recent but enormous development, leading to changes in food selection and later to supplements to ward off nutritional deficiencies, such as scurvy, rickets, and anemia. There were the revolutions in heating and cooling foods—cooking, baking, broiling, refrigerating, and freezing. The green revolution gave us the era of high yield crops through use of fertilizer and nitrogen, and there were innovations in harvesting crops with threshers and combines. Today we debate the soundness of genetically modified organisms. With this rich history of agricultural innovation, is it so impossible that we'd come to grow meat in a lab?

Even so, I have to confess that for purely personal reasons, the lab meat tasting involved a small dilemma. I'd been a vegan for three decades, and somehow this sample offered to me felt like a breach of that discipline. Would I still be a vegan if I ate the stuff? Even though the lab meat involved no suffering whatsoever for any animal—and indeed is meant to eliminate vast suffering—it still didn't feel right, and I had to think it through.

An analogy here would be egg consumption. I've abstained from eating free-range eggs largely because of my own identity as a vegan, as a long-established habit, and as a general protest against the broader system of industrial agriculture that mistreats animals so systematically. I confess that I don't have a moral question with anyone eating an egg from a well-treated hen—she lays the egg without pain, and it's as normal a biological function as a rooster crowing at sunup. Lab meat would seem to fall into that same category of harmless, cruelty-free protein.

"The vegan definition may be blurred, since it's an animal product," as Forgacs explains it. "But if you have a product that is .001 percent animal and the rest of it is derived from plants or brewed from plants, then you are largely eating a plant-based product." The distinguishing feature of the lab-grown meat was that it wasn't ever a being. There's no brain, no consciousness, and no pain. This was a big contrast with some of the bioengineering going on in the conventional meat industry. Some of its scientists have been trying to eliminate the "stress gene" in pigs. The idea is to make them less afraid of pain and death—a truly insidious project that would permit these creatures to be endlessly exploited, while losing even the dignity of being the conscious, feeling animals they are.

Despite my mild misgivings, I took a bite of the cultured meat chip. The barbeque sauce registered with me, but there was no awakening of my taste buds. In all fairness, though, even real beef jerky wouldn't do much for me either. The test will come, I thought to myself, when chefs and amateur cooks experiment with lab meat, and determine if it can perform in the pan or under the broiler like the meat they're used to. Perhaps the biggest question that remains is how the product will improve in taste and function as the company scales up, and we'll only find out when Forgacs' company grows and is able to put a rib eye—and then the whole, vast meat locker variety of foods—in front of dedicated carnivores to see if they clean their plates.

A year before I met Forgacs, I had attended the annual conference in New York of the Clinton Global Initiative. Each year, the Initiative hosts a competition for the Hult Prize, the world's largest student competition focused on the creation of new social enterprises. Teams of students go before judges with an eight-minute presentation on social entrepreneurship to demonstrate creativity, social impact, and economic viability. In 2013, the winning team came from McGill University, in Montreal, and its social enterprise, named Aspire, promoted insect farming in "countries with established histories of entomophagy." The team developed and distributed "affordable and sustainable insect farming technologies," after concluding that food insecurity was not just a matter of insufficient calories, but the wrong foods. "Not only do our durable farming units create income stability for rural farmers, they have a wider social impact by lowering the price of edible insects. This is central to our mission of increasing access to highly nutritious edible insects amongst the poorest, and therefore neediest, members of society." Insect-based food items are unlikely to find shelf space at Whole Foods or Safeway, but it is a potent idea for much of the developing world where eating crickets or beetles can pass for normal fare.

In delivering the award, President Bill Clinton joked about what his vegan friends would think of this project. He didn't seem worried about its threat to his own semi-veganism, since he wouldn't be sampling any of Aspire's prize-winning fare. But he wondered aloud what our traditional definitions mean in a world of innovation and problem solving around food.

The insect-farming project seemed to sacrifice palatability for practicality—putting a premium on delivering nutrients and protein in an affordable way. That was the same rationale for a new product called Soylent, whose creator, Rob Rhinehart, had a vision to nourish himself on the cheap and to avoid the "time and hassle" of eating. He developed a mix of foods, almost exclusively plants, and blended them together into a single shake that he says is nutritionally complete. A month's worth of Soylent, consumed twice a day, costs about $70—roughly the price of a single trip to the supermarket.

Whereas some food revolutionaries like Forgacs are working to duplicate the foods we love, Rhinehart is asking us to break free of our food habits, to love food less, and to adopt a more ascetic eating regime. He reduces food to a series of practical concerns centering on cost, time, and nutrition. Rhinehart wants people to reclaim the time, money, and energy they expend on preparing food, going out or traveling to eat it, savoring it, and subsidizing its often complex and elaborate production and presentation. For Rhinehart, it's all about the end game—about food as fuel. While many food critics have given his product harsh reviews—mostly because it's a challenge to the delights of eating and the enormous industries built around those pleasures—the company has already been valued at $100 million. That may reflect the company's potential to sell its product in developing countries struggling with food insecurity. Or perhaps it's a niche market in the United States where there's a class of consumers for whom convenience trumps all other concerns. There may also be a segment of people so desperate for an alternative to conventional agriculture that they're willing to do just about anything to break free of it.

Compared to insects and Soylent, Forgacs seems like he's got a relative advantage with his meat, because he's not asking anyone to forgo a conventional meal. The same goes for his leather operation. Not content to upend the eating of animals, Modern Meadow wants to disrupt the wearing of them, too.

The $60 billion leather industry processes twenty billion square feet of leather a year, and it has its own issues with toxins, pollutants, and other environmental hazards. Most of the leather comes from the hides of cows, of course, so it's a critical profit center for the cattle industry, and helps to subsidize its operations. There's also the exotic leather industry, a fast-growing sector that farms and kills alligators, crocodiles, and other wild animals for their hides.

"We are not going for perfectly identical," Forgacs tells me. "We are creating something in the case of leather that has design or performance advantages, just as with cultured meat, which can be healthier, more food safe, convenient, and involves no harming of animals or the environment." He says collagen is the main building block of leather, and he wants to create durable materials that stand up to abrasion. He creates dermis by growing skin cells, forming sheets, and layering them isometrically. "It's hide without hair, without flesh, without fat, and without all of the dirty stuff that tanneries work with." Forgacs says that his process for making leather uses fewer chemicals and half as much energy as the traditional method.

I asked him why go to all the trouble to make meat and leather from cells, when there are plant-based foods and synthetic leathers. For instance, the clothing designer Stella McCartney, daughter of Sir Paul, has a beautiful line of non-leather shoes and handbags. Forgacs told me that he knows that there are good products out there, but they just won't work for some people. "It's all fraught with so much history and sociology and notions of status," he observed. "It's so difficult to drive change person by person, and putting these products out in the marketplace, eventually in a seamless sort of way, has the potential to drive behavioral changes on a planetary scale."

In vitro meat and leather are subversive ideas, calling for changes throughout the whole system of production. But will they be any less fantastic than factory farming—which involves putting millions of animals, all meant to move, in cages in windowless buildings and altering their physiology to make them fast-growing meat machines? In nature, digested grass produces meat and leather, and it seems reasonable that plants can also provide the fuel for the growth of animal products in a controlled setting. If you can duplicate animal tissues used for meat and leather, and do it without all of the moral costs and the waste, isn't there a moral imperative to do so? If it's a functional equivalent but a superior process—"a dinner without a death" in a phrasing from Matthew Scully—what is the case against it? In effect, cultured meat removes, once and for all, any plausible claim of necessity from what all along much of humanity has viewed as the necessary evil of slaughtering animals for food. When Forgacs' enterprise, and others competing for success in these futuristic endeavors, displace the conventional meat industry, they will become case studies in the economic concept of creative destruction—destroying the bad and creating the good in its place.

Planting a Steak in the Ground

"I will eat this meat. I am man," declares the actor in a well-worn Burger King television advertisement—a male pride rendition of Helen Reddy's 1970s anthem "I Am Woman." The twenty-something guy springs from his seat at some unnamed white tablecloth restaurant, where he's on a date with his wife or girlfriend, after a waiter serves them two plates bearing just a few wispy vegetables. While the woman presumably is satisfied with such delicate fare, he isn't. Without even so much as "goodbye, darling" or even a departing wave, he stalks off—akin to the man leaving the women and children behind at the village and venturing off to the hunt that will result in real food for him and his family. "I am way too hungry to settle for chick food," he offers as he exits the restaurant.

He walks into the streets of the city—the proverbial jungle, a symbol of the wild—and declares "I am man, hear me roar." With his animal instincts now expressed, he says goodbye to civilization and all things feminine. He's "starved" and "incorrigible" and "on the prowl . . . for a Texas double whopper." He will "wave tofu bye-bye" because only meat can satisfy a hungry man. As dozens of men gather behind him, all possessing the same mad-as-hell-and-not-going-to-take-this-anymore attitude about vegetables and other feminized and domesticated offerings, they are collectively unleashed—and a bit unhinged, as they lift a minivan and dump it into the water from a bridge. The tribe is now a posse, better yet an army, displaying a second male instinct—the urge to go to battle. In this case, their adversary is a world that tells them to no longer be men. This army can only be strong and prepared for battle with meat in the belly. In this case, they are armed with double whoppers, and like animals tearing into flesh, they bite into their giant double whoppers as they march—no utensils, napkins, plates, or any of the other trappings of modern, feminized society. Meat has made them whole again, as the narrator intones, "Eat like a man, man."

There is an endless stream of advertising from food retailers equating meat with masculine strength—from Arby's celebrating its "Mega Meat Stacks" to Oscar Mayer invoking the eighteenth-century patriot, Paul Revere, who reminds viewers that he didn't make his legendary ride from "Boston to Concord while nibbling a quinoa-crusted protein bar." One McDonald's commercial simply features a thirty-second vertical panning of a towering Big Mac, while the narrator warns, "All vegetarians, foodies, and gastronauts, kindly avert your eyes." The notion is that the softhearted vegetarians or epicureans cannot bear the image of cooked flesh, but meat eaters cannot take their eyes off it. "You cannot get juiciness like this from soy or quinoa," the narrator informs us. Calling the viewer's attention to the iceberg lettuce, the narrator implicitly reminds us to keep lettuce in its place. It's an adornment, a garnish, and never to be trusted at the center of the plate. The burger "needs no introduction" because it's so obviously what we desire. For most of those in the meat industry, it's not enough to celebrate meat—one has to denigrate the alternatives, whether complaining about the emptiness of lettuce or the elitism of trendy foods like quinoa and kale.

Since Ray Kroc sparked a transformation of McDonald's in the 1950s, the company has delivered protein—along with fries and a soda—with speed, consistency, and low prices. In fact, when I was a kid, you could get your meal under the golden arches for a buck. It remains inexpensive today, with its cheap ground beef and other simple ingredients and its low-wage labor. And for any kids like me who were concerned about animals, McDonald's was happy to have us believe that our food came from "a hamburger patch"—with the burgers dressed up like dancing plants. Kids know it's not for real, but at least it changes the subject. The theme is Happy Meals, with a clown as tour guide, far from the melancholy or pity that arises—in children, especially—from the awareness of sad, suffering animals. The neat, colorful packaging and the take-away toys only increase emotional distance from the factory farm, the feedlot, and the slaughterhouse.

But with more than a third of American kids today considered obese, consumers are challenging the long-dominant narrative of the fast-food companies. Parents no longer accept such blatant manipulations of children, and are increasingly seeing fast-food companies as the symbols and agents of a nationwide epidemic of unhealthy eating. In the 2004 documentary *Super Size Me*, Morgan Spurlock famously ate three meals a day at McDonald's for a month, and gained twenty-five pounds after just three weeks. At day twenty-one, he experienced heart palpitations and his internist advised him to cease the experiment or face long-term health effects. In response partly to the public outcry created by *Super Size Me*, Eric Schlosser's *Fast Food Nation*, and Michael Pollan's *The Omnivore's Dilemma*, fast-food chains added salads and fruits to their short menus. But McDonald's made a change at the top at the end of 2014, amid lagging sales and a buzz about competitors, such as Chipotle and Shake Shack, offering plant-based and more humanely produced foods and setting up the purchasing experience so that customers can have a say in building the wrap or burger they buy.

Yet even as they face these challenges to their business models, some fast-food chains are doubling down on meat as part of their sales strategy. Wendy's, Carl's Jr., and others are offering bigger patties and slopping bacon on top of beef. Since 1950, Americans have increased their consumption of meat and fish by almost 50 percent—from 150 pounds per year to 220 pounds today. Americans eat more meat per capita than just about any people in the world—more than the Russians, the Argentinians, or the Australians. And that's perhaps why Andras Forgacs thinks the only way to upend the meat industry is to give people meat.

But there is a growing class of entrepreneurs who think there may be other practical ways to duplicate the taste and texture of meat, so that consumers can do something better for themselves, animals, and the planet—without any feeling of sacrifice or deprivation.

Ethan Brown wants to create the perfect plant-based burger, but he's nobody's idea of a hippie vegetarian. The forty-four-year-old does live in California, but he's got short hair parted on the side, a baritone voice, and the 6′5″, 220-pound frame of the former college basketball forward he was at Connecticut College. After working in the environmental field on fuel cell projects, he got an MBA at Columbia University. By the time he'd reached his late thirties, he'd found his calling and began seeking investors for a start-up company with a mission as direct as its name: Beyond Meat. Ever since he was a kid, having spent a good bit of his childhood on a farm in rural Maryland, he wanted to help animals. Here was a way to do that while reducing greenhouse gas emissions and promoting healthier lifestyles—all by giving people a satisfying and familiar option in the marketplace.

I went to see Brown in El Segundo, an industrial and residential town just south of Los Angeles. There are no bright lights or big signs on the building, and this is definitely not a food sales outlet or even a production facility. It's Brown's business and R&D center: a single-level brick warehouse. A small, hand-written sign on the front door directs visitors to enter through a side entrance, so I round the corner of the building and slip inside a propped-open door and walk right in, on the lookout for a receptionist. If they're worried about anyone stealing their secret recipes, they sure don't show it. I pass by a number of folks, and only by the time I am ten steps in do I finally manage to catch someone's eye and ask for Ethan Brown.

"Wayne!" Brown exclaims, "Great to see you!" He gives me a firm handshake and puts his other arm on my shoulder. Brown then leads me through his obstacle course of an office, where the furniture looks as if it's been moved around by a low-impact earthquake. The place is filled with metal desks, about half of them occupied, in a large open room with a high ceiling. No suits, ties, or dresses—it's all jeans and t-shirts. In the center of the room, there's a big piece of heavy equipment—which I later learn is an extruder and called "the Steer"—that doesn't seem to be doing much of anything, even though it's the company's alchemy machine. We walk through the kitchen, which in this business is not just a place where employees gather for lunch, but the creative center. Inside, master chef Dave Anderson works with a team of plant chemists, structural biologists, and other scientists to find plants that have lots of protein that can be fashioned into a modern meat replacement consumer product—not just beyond meat, but better than meat.

Eventually, Brown and I settle into a conference room where the table takes up 80 percent of the space, and I feel like we are nearly pinned against perpendicular walls. He reminds me that we'd met in my Washington, D.C., office in 2008. At that time, he tells me, he had big ambitions, but hadn't yet started Beyond Meat, which he founded a year later. Before that, he'd "been traveling around the mid-Atlantic region, where I was from, to sell veggie beef strips from Taiwan," he told me. "The Buddhist monks have been making these products for ages because you cannot have meat in the temples. They use soy and gluten, and they put the product through a low-pressure extrusion process to bind it together and give it a consistency and form that resembles meat. Good stuff, but not ready for prime time in terms of mass consumerism." Indeed, go to any Asian restaurant, and you'll typically find plenty of plant-based entrees.

"I wanted to find some scientists to crack the code" of realistic meat substitutes, he explains. "I started to work with people from the University of Missouri in Columbia. These scientists were taking plant products but using the extrusion process in a distinctive way—heating and cooling and reengineering the structure of these plants and blending them together to get something that was much more like meat in form and texture." Extrusion is a process already widely used in the food industry to make pastas, cereals, and a wide range of other products. "It's like a lock on a safe, and they were figuring it out," Brown says.

"What is meat?" he asks me rhetorically. "It's amino acids, carbohydrates, lipids, minerals, and water, and all are available in the plant kingdom. Why couldn't you create that exact same structure with these building blocks? We do in three minutes what it takes a cow to do in two years." He's confident that his process is not only going to work—he finds it impossible to imagine that there won't be a food revolution. "I want to beat meat in every way—taste, price, nutritional content," he tells me. "Look at Tesla. Its automobile design addresses environmental problems, but it's more than that. It's cool and it's elegant, even apart from its environmental benefits." This point is central to his vision—it isn't enough to appeal to environmentally conscious or animal-welfare-friendly customers. He is determined to produce a food that has stand-alone appeal and cachet.

It's all very experimental, but then again, if you think about it, so is the system that Beyond Meat and similar enterprises are trying to replace. Factory farming has been a fifty-year process of pushing past old boundaries, testing the limits of new techniques and also of the animals themselves. In the post–World War II period the middle class swelled, people flocked to the newly created suburbs, and manufacturing and technology surged. It was no surprise that the thinking behind the growth of so many forms of manufacturing would also affect agriculture—especially in light of a recent history of food insecurity. There had been massive farm bankruptcies in the 1920s, and even more in the 1930s, with the Great Depression and the Dust Bowl. World War II had brought rationing, as those on the home front were asked to plant "victory gardens" to deal with food shortages. The postwar era was a period of plenty and the emerging goal for farming, Ezra Taft Benson, President Eisenhower's secretary of agriculture, famously declared, was "to get big or get out."

With the US government pushing a food pyramid for decades that extolled meat as a source of vigor and strength, fruits and vegetables were bunched into a single food group and shunted to the side of the plate. In the 1970s and 1980s, the government worked with the meat industry to create check-off programs, pouring tens of millions a year into consumer marketing programs for animal-based commodities. The best known of these programs produced such slogans as "Beef: It's What's for Dinner," the "Incredible Edible Egg," "Pork: The Other White Meat," and "Got Milk?" Consumption of milk, eggs, and chicken surged, while beef and pork more than held their own even as health concerns arose about their links to heart disease and cancer. An international panel of experts convened by the World Health Organization announced in October 2015, to great skepticism from the public, that eating processed meat like hot dogs and bacon raises the risk of colon cancer.

Yet, there's a lot to unwind—with decades of government and industry promotion of meat. In 1964, Congress created the US Meat Animal Research Center, which is now a sprawling fifty-five-acre campus in western Nebraska. Researchers there and at more than forty other research facilities administered by the Agricultural Research Service, along with many more land grant colleges, study how to make the meat business even more efficient, mainly through genetic and reproductive manipulation of animals. They've conducted experiments to select for fast-growing breeds of farm animals; confine them in buildings; dose them with beta-agonists, steroids, and antibiotics; and control the feed, temperature, and every other aspect of the animals' lives.

In early 2015, *The New York Times* broke a story, informed by a veterinarian and whistleblower at the Nebraska Center, James Keen. When I sat down with Dr. Keen a week later, he described to me a kind of dystopian atmosphere at the federally funded facility. Researchers had worked to increase litter sizes for pigs and cows, with the result that the mother pigs crushed their piglets and the calves were born deformed. They'd even tried to develop "easy care" sheep, only to see them attacked by coyotes or battered by hail without the shelters and shepherds that some researcher thought they could do without. The government spends more than $1 billion a year on this broad class of experiments, and on top of that the Nebraska Center raises and sells thousands of animals a year for slaughter—supplementing its budget by essentially getting the government into the meat business.

Chickens now go to market after just less than seven weeks—down from twelve weeks seventy years ago—reaching "market weight" in half the time once needed. What's more, their market weight is four times what it used to be. According to researchers at the University of Arkansas, if humans grew as fast and as large as chickens, a 6.6-pound newborn baby would weigh 660 pounds after two months. Some chickens cannot even stand—the industry has a special term for them, "wing walkers," since they try to drag themselves along the ground by using their wings. Turkeys are bred to be grossly obese, chronically lame caricatures of the fast-flying wild birds from whom they descend. Cows at industrial dairies now yield an average of 27,000 pounds of milk a year, so far beyond what they would naturally produce for their calves that these mothers are typically spent—used up and unable to go on—when they are just three to five years old.

Like the freakishly large birds and animals on factory farms, hyper-intensive agriculture in America has also selected for a particular kind of farmer—the big, fast-growing kind—while crowding out farmers with a sense of limits and a traditional commitment to animal care. To gain a competitive advantage, these people took on debt to acquire bigger buildings and ever more animals to fill them with. Those who didn't employ these techniques had a rough go of it. Many went bankrupt, especially during the 1980s, as major companies vertically integrated their operations. Meat industry leaders, such as Tyson and Smithfield and IBP, not only own the animals but also set prices by controlling slaughterhouses. Within the last forty years, there's been a 95 percent reduction in the number of egg producers, a 91 percent loss in the number of pig producers, and an 88 percent decline in the number of dairies. Yet total animal production in the US has steadily increased—from about 1.5 billion animals in 1960 to 9 billion today. Vastly more animals are farmed by fewer and fewer companies, many of which are employing harsher and harsher methods. The family farms that remain are highly vulnerable, because tens of thousands of them are contract growers working for conglomerates like Tyson Foods and Seaboard Foods. These farmers assume the costs and debt in acquiring large buildings and physical infrastructure, but they do not own the animals—they are just housing them, with the prices set by the processors.

More than ever, all the rest of us are deeply disconnected, not only from these economic dynamics, but also from the fate of the animals. And that suits the meat industry just fine. Removing any thought of the handling and killing from our daily experience, and presenting a small part of an animal in a package at the supermarket, sanitizes the experience and removes at least some of the moral discomfort. Naturally, the last thing the industry wants is for you and me to think, even momentarily, of all that happened before the meat reached the grocery store. The ideal producer, by their lights, is someone who doesn't let conscience get in the way of what he or she wants. And that's their image of the ideal customer too.

The agribusiness companies even take it one step further. On their packaging, they stamp pictures of family farms, or they invoke words like "natural" or "humanely raised" that have no connection to reality. Yet they try to stop investigations by animal-protection groups inside their facilities, pushing "ag-gag" laws to criminalize the mere investigation, or even photographing, of the conditions at factory farms. They have also tried to enshrine their current practices as legal by passing "right to farm" measures intended to short-circuit reform efforts. In recent years, unfortunately, several states have adopted such laws, and some have even enshrined this free pass for agribusiness as a constitutional protection.

These efforts to perpetuate a fiction about industrial agriculture or to enshrine these practices in law, while skating past the reality of ruthless treatment of animals, have set the stage for the new round of innovation in food and agriculture. Science, in a sense, got us into all this trouble, but maybe science can get us out—or at least that's the hope of people like Ethan Brown and Andras Forgacs.

Brown's Beyond Chicken has greater short-term market potential than Forgacs' Modern Meadow, partly because he's attempting to improve on a process that has been around, in some form, for thousands of years. In fact, he is already well along in the marketplace—I regularly get my Beyond Meat "chicken" strips at Whole Foods, and you can find it as well at chains like Safeway and Kroger. On a recent anti-dogfighting deployment, I dashed into a Tropical Smoothie and found Beyond Meat chicken on the menu in Dothan, Alabama—not the sort of place that comes to mind for ready vegetarian fare. It was delicious, and I knew it had all the protein I'd need. I remembered that Brown was featured on NBC's *Today* show, and the program did a blind taste test, with Beyond Meat chicken and ground beef going up right against the real stuff. If there was a difference, it was lost on the four hosts of the show.

The HSUS had been so impressed with Beyond Meat's business plan that it put half a million dollars into the company, as a double bottom-line investment. In investing in a social enterprise, we seek a financial return, but also invest so that the company can do work that's beneficial for animals and for the whole of society. With the average American eating about twenty-nine land animals a year—mostly chickens—that's a savings of a different kind. For the HSUS and for Brown, it's tackling the problems of factory farming and unsustainable meat consumption from an angle we'd not pursued before—by working to compete on the production side to give consumers more choices. It's a strategy that doesn't require winning over consumers on moral or ecological reasoning—which simply may not be persuasive for some folks. Rather, the goal is to appeal to their gut-level instincts on taste and desire, by giving them something loaded with protein and familiar in texture.

Brown was not the first mover, by any means, in this category. Beyond Meat is just one of dozens of companies that have developed non-animal protein foods in response to consumer demand for low-fat, no-cholesterol, no-cruelty products. Gardenburger, founded by Paul Wenner in the early 1980s, was the original mainstream brand; it's now owned by Kellogg and found in supermarkets throughout the nation. It competes with Yves, Quorn, Boca, Tofurky, Field Roast, Garden Protein International, MATCH, and MorningStar Farms, which is also owned by Kellogg and is the largest vegetarian food producer in the United States. In 2014, Pinnacle Foods paid $174 million to acquire Garden Protein International, whose Gardein chicken strips and tenders, ground beef, and fish fillets are available in twenty-two thousand stores. Fast-food vegan chains such as Veggie Grill—with dozens of outlets in major cities—have Gardein products on their menus, featuring them in salads, wraps, and sandwiches. So does Lyfe Kitchen, founded by a former McDonald's executive, and offering plant-based foods and more humanely raised animal products at its fourteen outlets. Chipotle, one of the fast-food industry's fastest growing players (with nearly two thousand outlets and 2014 revenues of $4.1 billion), makes a spicy tofu-based product called Sofritas alongside its more humanely sourced animal products. Even traditional burger joints, like White Castle and Burger King, have vegan burger offerings now.

While the growing roster of companies making and offering plant-based proteins is an indicator of changing times and tastes, the sector is still tiny compared to the meat industry, which generates hundreds of billions in sales. Ethan Brown wants to see Beyond Meat in the meat case, going head to head with chicken and beef on price and taste—since the availability of an inexpensive food product, rich in both flavor and protein, has always been its prime selling point. "We have been eating meat for two million years," Brown tells me. "We are the way we are because of meat consumption. Meat allows us to develop this big brain of ours, but it wasn't like we were after meat for meat's sake. We were after nutrient-dense food. What if we create a more nutrient-dense food that tastes just as good as the meat we've craved so long?"

That craving, odd as it seems to people who truly get beyond meat, shouldn't be underestimated. In 2014, a study by the Humane Research Council revealed that 84 percent of people who at one point identified themselves as vegetarians later backtracked and resumed some amount of meat eating, albeit at much lower quantities than is typical in America. There's no single explanation for this sort of mass retreat, but lapsed vegetarians cited a combination of cravings, the lack of good plant-based products, and social pressures as primary factors. Brown's aim is not to create a vegetarian culture, but to reduce the number of animals raised and slaughtered for food overall, by giving people more options. He wants not just to hold current vegetarians in place, but also "to bring new people into the category." His breakout product, which he had just designed with his team when I visited, is the Beast Burger—made of a combination of yellow peas, beets, and other plants.

The Beast Burger was to have its marketplace debut in Whole Foods in January 2015, and at Brown's office I was going to get an advance tasting. As I readied to bite into both products, Brown told me they don't have any supplements—except what sunshine and soil infused in them. "The burger has only 250 calories and it has only good fat—the kind you need. It has more protein than beef, more omegas than salmon, and it has calcium and anti-oxidants." And unlike conventional ground beef, it lacks steroids, hormones, antibiotics, and other additives that aren't doing your system any favors.

Brown has competition in the race to develop meat facsimiles that replicate the look and taste of meat. Dr. Pat Brown, a scientist at Stanford University and no relation, is also engineering plants to replicate meat through his company Impossible Foods. A biochemist, Dr. Brown is finding ways to develop substitutes that look, smell, and taste exactly like meat—and the key to achieving that look-alike and taste-alike quality is "heme," a protein found in nodules attached to the roots of nitrogen-fixing plants that is similar to myoglobin and hemoglobin (which make blood look red). In July 2015, Google reportedly offered to buy Impossible Foods for $200–300 million. Like Ethan Brown, the Stanford scientist has been fabulously successful in attracting investors, with a second round of $108 million in capital announced in October 2015. With about one hundred employees and a remarkable focus on the science of building a duplicate of meat from plant building blocks, he's determined to get closer in form to animal meat than any prior product in the marketplace. "There are limits to what a cow can be—a cow can only be a cow," he told *The Wall Street Journal*. "We can make anything."

"Plant-based eating is going to be the future," said Barb Stuckey of the food product development company Mattson, "but you have to be able to deliver the craveability you get from meat and dairy, the savory, crumbly, delicious, sensory experience." As I scarfed down my Beast Burger, I sensed that the technologists and entrepreneurs were getting closer with each fabrication. The burger was juicy and rich, and it popped with taste. "You can walk right out of the gym and run," Ethan Brown added. "You feel better after you eat it. Why *wouldn't* you have this?"

In October 2015, Brown announced a new board member for his company—none other than Don Thompson, the CEO of McDonald's let go in 2014 and who in 2012 worked with me and Carl Icahn on the policy to eliminate gestation crates from his supply chain. Remember, Brown was a team sports guy before he was an entrepreneur. He knows he needs a team approach to get his product into the meat case and into the diets of millions of regular Americans.

Egg Beaters

Perhaps only in Silicon Valley do people think about computing when there's talk of chips and cookies. But the edible kinds are now sneaking into business discussions in the high-tech world. Andras Forgacs has attracted a cadre of venture capitalists to invest in his in vitro meat chips. And in May 2012, none other than Bill Gates, the world's wealthiest man and benefactor of its largest foundation, and Tony Blair, the former British prime minister, came to northern California and ate some cookies with some very special ingredients.

They'd come to a resort in Sausalito, in Marin County, to see the companies they back through their involvement with the Menlo Park–based Khosla Ventures. Silicon Valley is now investing capital in new businesses intent on disrupting industrial agriculture and building humane, sustainable, and affordable solutions in a world expected to have two billion more mouths to feed by 2050—and Khosla Ventures is at the forefront. Led by the Indian-American billionaire Vinod Khosla, the firm is betting that a company called Hampton Creek is going to make the factory-farmed hen something of a rare bird.

Hampton Creek is mix-and-matching and then blending different types of plants and their proteins to mimic the functionality and taste of the common egg, with the goal of "taking the chicken out of the equation," according to chief executive Josh Tetrick. An Alabama native who was a linebacker at West Virginia University, Tetrick is now leading one of the fastest-growing businesses in the world. In asking Gates and Blair to compare cookies—the conventional kind made with eggs versus the egg-free type cooked up by Hampton Creek—Tetrick gave them the sort of assignment that any person with a sweet tooth could hardly turn down. "I can't tell the difference," Bill Gates exclaimed after he dug into the chocolate chip cookies at the summit event, while Blair predicted, "This will be big in Europe."

Battery cages hadn't just faced an enterprise threat from our cage-free campaign, but from a different kind of competitor in the marketplace. Just a few weeks after our trip to Marcus Rust's farm, Josh Balk and I made a special visit to Hampton Creek, a company that's developing common food products without eggs and other animal-based ingredients. While Balk has kept busy negotiating corporate animal welfare reforms for the HSUS, he also stayed focused on an objective of a perfect replacement for the egg. He helped recruit his long-time friend Josh Tetrick to run the company. They named the enterprise after Balk's boyhood dog, Hampton. "We added the word 'creek' because it sounded nice," he told me.

I wanted to see what Balk and Tetrick had been building, so we drove to the company's ninety-thousand-square-foot office and research facility in San Francisco. There, two master chefs had done some handiwork with a secret blend of plant products, a few elongated kitchen tools, and a pan and flame. It was 4 p.m.—a no-man's-land for the regimented eater—but they had already plated some of my favorite comfort foods. Resistance was futile.

Chris Jones and Ben Roche told me that they headed west from Chicago, where they left their gig at the famed gourmet restaurant Moto, because they were excited about Hampton Creek's mission. Balk told me that "no company has gotten its products into Walmart this fast"—taking just two years to go from its initial business plan to occupying shelf space at the nation's largest food seller. Its first commercial offering, "Just Mayo," a healthier, more affordable, and tastier version of mayonnaise, was already sold in thirty-three thousand stores in the United States and China. It was the top-selling mayonnaise in Whole Foods, and it was outselling the dominant brand, Hellman's, in some chains. "We used to cook for the 1 percent, and now we cook for 100 percent," said Jones, a classically trained chef who, in his gourmet cooking days, was featured on *Top Chef.*

Situated in the SoMa section of San Francisco, the façade of the Hampton Creek facility looks like a spit-shined former auto body shop. But the work here is about as unconventional as can be. The aim is to make obsolete an entire sector of industrialized animal agriculture. And the company's work doesn't end with blended products like mayo. As a first stage for the company, name any other product containing eggs, even scrambled eggs, and they want to find a superior replacement for it.

I asked Jones about the other plant-based egg substitutes in the marketplace. He seemed hesitant to compare existing egg replacers, or other plant-based substitutes. But I pushed, and he confessed that "our technology is better. They don't have the mouth feel. It has limited uses, and breaks down with heat. It's why we are in food service, as well as in Walmart, with a finished product. Our product holds together and has the taste and texture you expect from mayonnaise."

Jones and Roche had given me a primer on their products and also presented me with breakfast, lunch, dinner, and dessert—all at once. With three meals in front of me, I figured I'd keep them in order by starting with the lightly battered French toast, glazed with non-dairy butter and maple syrup. The French toast, Roche said, "is what sold me on the company."

I loved it, too, but the fettuccine with a creamy sauce demanded my attention as well. It tasted just right. Jones told me that fresh pasta had eggs in it, but ours has "30 percent more protein." A single egg producer named Michael Foods, Balk mentioned, uses thirty million eggs a year for its own pasta product. Given that the specially bred laying hen produces more than 260 eggs a year, "that's 115,000 hens removed from battery cages just for that one line of pasta. Imagine once we really get rolling how many animals we'll get out of cages."

"So much of cruelty is hidden," Balk said, as I finished off the fettuccine. "People are not going to know the difference. They are not going to care that there are no eggs in the product, and all of this suffering will be reduced."

"Polenta typically has butter and cream," Jones said as he previewed my dinner selection. "Our mayo gives it the creamy texture." I was getting full, but if I ate fast enough, I thought, I could stay a step ahead of my stomach. Soon there were cookies, cookie dough, and ice cream for dessert. I was reminded of a scene from Monty Python's *The Meaning of Life*, when a grossly obese customer looks at the menu and says, "I'll have the lot of it." I soon held up the white napkin, surrendering with a plea to save the rest for later. Hampton Creek already has thirty-six food products, with mayo the best known and the top seller, and the total number of offerings is sure to expand.

I was there at a heady time for the Hampton Creek team. The day before I arrived, the global food giant Unilever, maker of Hellman's mayonnaise, had withdrawn its lawsuit challenging Hampton Creek's Just Mayo product—backing away from the argument, grounded on US Food and Drug Administration regulations dating back to the 1950s, that the start-up could not call its product "Just Mayo" because it didn't have eggs in it. The multinational had faced a firestorm of criticism for filing the suit against a start-up, inadvertently giving a greater lift to Hampton Creek's fast-developing brand. To me, it was unfortunate because two years earlier Unilever had demonstrated a strong streak of social responsibility in making a commitment to switch entirely to cage-free eggs.

Even though Unilever had dropped its suit, the FDA wrote to Hampton Creek and threatened to revoke its use of the term "mayo." But just as the multinational corporation had retreated after making its original complaint, so, too, did the federal government. In December 2015, the FDA cleared the way for the company to continue to use the term, with just a few modifications to the label.

Only after fending off attacks from Unilever and the FDA did Hampton Creek come to learn where these attacks had originated. It had been in the crosshairs of a quasi-government entity called the American Egg Board (AEB), a USDA-authorized entity that acts as a promotional arm of the egg industry that is funded by taxing producers on their egg sales (transferring costs to consumers). Emails discovered through a Freedom of Information Act request found that the AEB conspired to thwart

Hampton Creek's initial efforts to get its products placed in grocery stores. Joanne Ivy, the chairwoman of AEB, wrote to the world's largest public relations company in August 2013 and suggested it "would be a good idea if Edelman looked at this product as a crisis and major threat to the future of the egg product business." Once these documents about a public relations attack on Hampton Creek by Edelman came to light in the fall of 2015, it wasn't long before Ivy resigned, since the federal program exists to promote the egg industry and not to denigrate its competitors. Among others, US Senator Mike Lee, a conservative Republican lawmaker from Utah, asked, "If these Great Depression era institutions have outlived their purpose, and if evidence suggests they behave like state-sponsored cartels that intimidate and handicap their competition, should Congress continue to authorize them?"

The very same day I visited, the team at Hampton Creek also had announced a series C investment round of $90 million. With twelve billionaires investing in the company, they had become the toast of Silicon Valley. In fact, investors had pledged $150 million, but Tetrick and Balk had turned away some of the money because they only allowed investors whose goal was permanent, systemic change in the global food industry, as opposed to a quick sale of Hampton Creek to a conglomerate. In addition to Khosla and his partners Gates and Blair, investors included Eduardo Saverin, co-founder of Facebook; Nicholas Pritzker of the Hyatt Hotel chain; Jerry Yang, former CEO of Yahoo; Peter Thiel, co-founder of PayPal; Marc Benioff, founder and CEO of Sales Force; Li Ka-shing, Asia's richest man with assets of more than $30 billion; and the Singaporean government.

Dan Zigmond, a scientist who came to Hampton Creek from Google and described himself as on the "dry" side of the business (before they start cooking the products), said the start-up is a "discovery company." He told me that there are 400,000 plant species, and each one has between forty thousand and fifty thousand proteins; this means there are eighteen billion plant proteins to choose from. He has a team of scientists studying the properties of these plants. "We are trying," he added, as he surveyed plant chemists and protein scientists hunched over centrifuges and computers, "to find where the valuable proteins are" and to replicate the best characteristics of eggs. "We've gone through five thousand to six thousand samples, but we think there's even more out there," he said. "We've only just scratched the surface. The scramble is the big new product—the moon shot that we are working on."

As with Modern Meadow and Beyond Meat, Hampton Creek invests in applied science, an approach that blends hard science with invention and testing in the kitchen. "We take these ingredients and bake them into model systems, into a muffin, or pound cake, and then we run a series of tests on those," Zigmond told me. "We measure the height and volume of the product to understand if it's rising, holding together like an egg would." But it has to be functional. "Is it a good emulsifier?" he asked. "Is it a good aerator; is it a good gelling agent? We are using data and data science to direct that search. We take all the data and try to narrow and direct our search more."

Perhaps no one has studied eggs so closely, and from as many angles, as the team of scientists at Hampton Creek, who aim to replace them not only in America but everywhere. "We want food products to last a long time," explained Lee Chae, who leads the science team, and that's accomplished by testing the plant proteins and seeing if they hold together and have the right composition in the pan and on the plate. "If we want to go into emerging markets and help people who don't have refrigeration, who don't have access to clean water, it must have long shelf life."

Tetrick told me the range of products that Hampton Creek can make in the long run, with its commitment to research and testing, is practically limitless. Mayonnaise is the number one condiment in the United States, and the Unilever lawsuit and the American Egg Board campaign were strong signs that the innovators at Hampton Creek were already disrupting that market. "We are beating eggs on every count—on price, on protein, on health, and definitely on animal cruelty," Tetrick said. "We can improve on it, and make it that much better."

The folks at Hampton Creek are dazzling entrepreneurs and scientists, people of real talent and vision. And while many consumers love the taste of eggs, as they do beef and pork, there are so many eggs built in as "hidden" ingredients that consumers just won't feel cheated if the eggs are replaced. Sure enough, in March 2015, Hampton Creek announced a deal with Food Buy, which spends $18 billion on foods per year as the purchasing arm of Compass Group, the world's largest food-service provider. The company was ditching its egg-based mayo, cookies, dressings, and other items to be replaced by Hampton Creek's product lines. "We love being in Target and Walmart and Whole Foods," Tetrick said, "but this is a world of head-turning scale, with flows of food through this system that are just enormous." In the fall of 2015, Tetrick attracted another round of private investments—$110 million—and the valuation of the company approached a billion dollars.

Changing the way we eat—substituting more plant-based products for animal-derived foods—would drive a number of changes for the good. Most Americans are deeply detached from the unending misery that hundreds of millions of hens endure on battery cage farms and at slaughter every year. There's the debeaking of birds, so they don't hurt their cage-mates by pecking. Then there is the issue of the male chicks. Of no use to the industry, they are stuffed in bags and suffocated, or they are macerated—ground up alive just after birth. The industry is trying to find a way, Chad Gregory told me, to identify the sex of the birds before they hatch so they can avoid killing them, but until then, one bird must be ground up alive for every one that lives—a moral math that no one would ever volunteer to explain, much less to justify. Indeed, in March 2015, University of Leipzig researcher Maria-Elisabeth Krautwald-Junghanns announced she'd developed a way to determine the sex of a chicken inside the egg just two to three days after incubation. The German agriculture minister said the new method will be used on all farms in the country by 2017. This would solve the problem of waiting until hatching and then killing half of all the new chicks. That's about forty-five million baby male chicks a year in Germany, and perhaps a billion worldwide—a merciful innovation that could solve one of the ugly, hidden horrors of egg production. (The German Poultry Association also announced an end to beak cutting for laying hens and turkeys—a development that should spread to other nations once producers in other nations see it work.)

When I visited with him in May 2015, Chad Gregory was also dealing with another big and very ugly problem—an avian influenza outbreak that had started just two weeks before we visited and that had spread suffering and death, along with plenty of angst from producers. Bird flu jumped from one industrial farm to the next, and whether the flock had an infection rate of 1 or 100 percent, state and federal authorities ordered entire flocks destroyed, and created a red zone around them where any movement of animals was forbidden. Small farms, and especially outdoor flocks, went almost untouched by bird flu, with the virus not able to survive as well outside and the birds possessing better natural immunity from being outside and in overall better health. The big confinement farms were vulnerable because they massed so much life into one space. In the case of an outbreak of a highly infectious disease, they all had to go. Three million birds here, five million there, and the body count becomes unfathomably large pretty quick. "There is an inverse relationship between accelerated growth and disease resistance, which means faster-growing birds are more susceptible to illness," according to Michael Greger, M.D., director of public health and animal agriculture for the HSUS. Three months after the first farmers reporting infection at a major battery cage facility, there were more than forty-eight million birds dead from the virus, or gassed by authorities—more than 90 percent of them laying hens. That was more than 40 percent of Iowa's hen population and about 15 percent of the national flock, and farms owned by Marcus Rust and Dave Rettig were among those affected. It was an awful circumstance for the birds, and Chad Gregory told me that guys he'd never seen cry had found themselves weeping.

Six weeks after the avian influenza outbreak, the price of a dozen eggs had doubled in most parts of the country. Whataburger Restaurants, which has nearly eight hundred restaurants in ten states, shortened its breakfast hours, and many other food retailers were themselves scrambling to find eggs. For the first time ever, the United States was talking about importing eggs. At the time of our meeting, the US government had already handed over its first grant, of $309 million, to egg farmers to compensate them for their losses and the total outlay for the US government would be a billion dollars—yet another hidden cost to the American consumer. Agricultural economists estimated the impact of bird flu just in Iowa and Minnesota would be an additional $1 billion to the local economy, with lost business for feed sellers, transporters, and others tied to the egg industry. It was also an ugly way for the birds to go, gassed in portable chambers or suffocated by water-based foam that is pumped in and swamps the birds. According to Dr. Michael Blackwell, the HSUS's chief veterinarian, it can take three to seven minutes for the birds to die. Foam doesn't work for the birds stacked high in battery cages, however, and the USDA in September 2015 issued a statement supporting in some cases shutting off the vents and heating up the houses so the birds suffocate—an unthinkable and even more miserable way to go.

Despite these government payouts, and perhaps because of the mass killing of the infected flocks, the market was reshaping the landscape right before our eyes. Josh Balk told me the major food retailers didn't like what they were seeing. "The bird flu outbreak caused so much supply chain disruption that Hampton Creek was bombarded with calls from major food corporations that they want to move away from purchasing eggs as ingredients and phase out the sale of products that contain eggs," Balk recounted. In June 2015, 7-Eleven, the world's biggest convenience food store, swapped out egg-based mayonnaise for Just Mayo for all of its prepared sandwiches. Hampton Creek, to be sure, would never have to cope with avian influenza or with mass killing, or with the disposal of millions of animals, nor be a candidate to get a massive emergency federal payment. Tetrick added that the company was "growing pretty fast before this thing happened, but it's amplified everything across the board."

With animal agriculture, there are so many problems to solve; misery and death pervade the process in the highly industrialized operations. And beyond the suffering, there's no way around the basic problem that animals use up a lot of resources: they need lots of food to grow, inefficiently converting plants into eggs, meat, and milk. "By 2050, the world's population will grow to more than nine billion and our appetite for meat will grow along with it. The demand for meat will have doubled between 2000 and 2050," Bill Gates said in a blog post after his CEO summit. "That's why we need more options for producing meat without depleting our resources. Some exciting new companies are taking on this challenge. They are creating plant-based alternatives to chicken, ground beef, even eggs, that are produced more sustainably, and taste great."

Those are the words of a man with considerable means and quite a record in turning simple and complex ideas into world-changing innovation. It's a story we'll all see unfold for the rest of our lifetimes, as old systems are seen for what they are—cruel and entirely unnecessary—and humane alternatives rise to take their place. In case after case, as new products become healthier and better tasting, the choice will come into sharper focus. If nothing is lost except the cruelty, as Ethan Brown would ask, why wouldn't we choose the new over the old?

CHAPTER FOUR

Now, That's Entertainment

"Very early in my life, possibly because of the insatiable curiosity that was born in me, I came to dislike the performances of trained animals. It was my curiosity that spoiled for me this form of amusement, for I was led to seek behind the performance in order to learn how the performance was achieved. And what I found behind the brave show and glitter of performance was not nice."

—JACK LONDON, *Michael, Brother of Jerry*

CGI and the New Era of Film

In a scene dramatizing the theme of "honor among thieves"—from a 2002 episode of HBO's *The Sopranos*—mob boss Tony Soprano suspects one of his deputies, Ralph Cifaretto, has killed a racehorse they jointly owned. Observing Cifaretto's indifferent response to the violent death of Pie-O-My in a stable fire, Soprano assumes the $200,000 in insurance money provides the motive. Tony's in a rough business and hardly the sentimental type, but killing a horse—well, that's going too far. Things get worse from there, at least for Ralph, and the scene ends with Tony calling over another deputy to help clean up the aftermath.

In real life, a decade later, there was quite a different dust-up at HBO regarding a horse incident at a stable. Network executives cancelled one of HBO's most hyped shows, *Luck*—which dramatized the world of horse racing—in its second season after Real Awesome Jet reared up as she was being led from the set back to her stall, flipping head first onto the ground. It was the fourth death of a real horse on the set, and an embarrassment to the network that had long enjoyed an animal-friendly reputation.

About the same time, there were rumors of animal injuries and death during production of *The Hobbit: An Unexpected Journey*. Reportedly, twenty-seven animals perished. And a friend of mine in Washington, Chris Palmer, came out with a book exposing unethical conduct among nature filmmakers, who were using captive animals and passing them off as wild, while staging inauthentic stunts that put the animals at risk.

There have always been abuses of animals in filmmaking. Yet movies have also been a powerful means of dramatizing the bond between people and animals. These great stories can and should go on—but without the hidden story of cruelty behind them.

To get a closer look at all of this, I paid a visit to film director Darren Aronofsky. When I saw his film *Noah* in 2014, I was dazzled by what he'd done with new technology. It was an ancient tale, of course, drawn from the Book of Genesis. Yet Aronofsky's vision, and specifically his representation of the filling of the ark, was a pretty breathtaking and distinctive retelling.

With direction from the Divine, and just a week ahead of the deluge that would submerge the grasses and the forests and then the mountains, animals descended from the skies, pulled themselves out of burrows or dens, and climbed up the entry plank. Some familiar, others exotic and fanciful, they wore fur, feathers, and scales, sported bright colors and dazzling markings, spiked headwear, and elongated tails. Creatures of every kind streamed into the ark, each settling into a nook designed just for them. Inside the cavernous ship, the animals drew breath from a blend of oxygen and an intoxicating vapor that put them into a torpor that the sharpest swells would not undo in the weeks ahead. But before drifting into consciousness, the lions, the lambs, and all the others were at peace. There had been no scrapping—instead, a sort of unspoken covenant emerged among species. God had declared all the animals "good" after He created them. But men had defied Him. He was wrathful, and would now wash away all. Only the life nestled into the ark, two of every kind, would go on. The six-hundred-year-old Noah had built the ship, and now he had been entrusted with the most precious steerage any vessel ever held.

If Noah had faced the challenge of fitting all the animals into the ark, Aronofsky was met with the task of packing them into a replica made for his 2014 epic. Given the near-universal familiarity of Noah's story, Aronofsky knew he had to grab viewers with the ark-loading scene; he did so by filling the screen with a beautiful and diverse array of animals. "We just needed a lot of creatures," Aronofsky told me. It was a task worthy of Cecil B. DeMille, and yet Aronofsky pulled it off without a single animal cage or keeper on-site.

When I arrived at Aronofsky's office in the Williamsburg section of Brooklyn, I triple-checked the address. From outside, the building looked run-down, and the door was plastered with stickers. I told myself this could hardly be the workplace of the acclaimed director of *The Wrestler* and *Black Swan*.

After getting buzzed in, I climbed a few steps into Aronofsky's spacious, open-air office, with about a dozen people staring at their computer screens and some on their cell phones. Not for the last time that day, I was reminded that first looks can be deceiving. Casually dressed, with a stubbly beard, Aronofsky greeted me and we sat down to chat about the message and making of *Noah*. Not long into the writing and planning of the film, he told me, he realized that using real animals would be impossible. "The problem with captive animals is, you are dealing with a very small bandwidth of types of animals," he said. "What's in captivity are animals you'd find in a zoo, and the animal kingdom is much more complicated and rich and diverse, so I knew early on I didn't want a polar bear, the giraffe, and the elephant, and the stock animals we all know. I wanted to try to remind the audience that this was a much bigger miracle because the animal kingdom is so diverse."

There were other practical concerns. "There have been many, many stories where you try to bring live animals together, and it's a big problem," Aronofsky added. "You bring a predator out first, and the prey species will smell it, and that animal will freak out."

The story of Noah has fascinated Aronofsky since he was a boy. At age thirteen, before he'd had aspirations of moviemaking, he wrote poetry about the famous Noah character in the Bible. That poem won him an award in a national competition, and he was asked to read it in public in the early 1980s. "That experience, of being recognized for your creative work, pushed me in the direction of doing more writing and poetry, and ultimately sent me down my professional path."

The tale of the Ark is the first and best-known animal rescue story of all time, and that attracted Aronofsky. "Noah really is the first environmentalist," Aronofsky said, "and a lot of Christian groups have attacked that statement, but how else do you describe a guy who's there to save every animal of the world, two by two, with a boat that he spent ten years building? He's there to save Creation."

"At a minimum, Adam and Eve were vegetarian, probably vegan, because they are commanded just to eat from the garden and not to touch the animals," he told me. "But the reason that God destroyed the world is because of man's evil acts on the Earth. They had corrupted the Earth, and one of those acts is that they were probably eating live flesh," and that was in violation of the proscriptions of God. "It made sense to us that somehow a descendant of Seth might be somehow different because he is chosen by God and maybe he is following the path that is expected from God, and that's why we made Noah a vegetarian as well."

Aronofsky's background as a student of animal behavior and an amateur naturalist also attracted him to the story of Noah. And it certainly made him well suited for the immense challenge of recounting the story in an original way. During his teens he pleaded for an assignment from the School for Field Studies, which sponsors students to go abroad (or at least afar) to learn about ecology and wildlife. A sustained lobbying effort got an assignment to Kenya, where he worked "to prove that wild animals have better water strategies than domesticated animals."

"Then I went to Alaska to Prince William Sound, and this was two years before the *Exxon Valdez* spill," Aronofsky told me. "It was an incredibly pristine environment when I was there. We were out on kayaks and never saw any other humans for five weeks, living at the base of a glacier and studying thermoregulation in harbor seals." As an undergraduate at Harvard, Aronofsky took up filmmaking while earning his MFA at the American Film Institute Conservatory.

While filming his 2005 work *Fountain*, Aronofsky came face-to-face with the conflict between his love of animals and his love of film. For *Fountain*—a science fiction thriller starring Hugh Jackman and Rachel Weisz, to whom Aronofsky was briefly married—the director ordered up live monkeys for a scene. The source was an exotic animal trainer, who transported them in the back of a truck—in a cross-country drive from Vancouver to the shooting location in Montreal. Before that, Aronofsky had given little thought to where the wild animals used in movies came from. "It was awful," he recalled. "I saw the cages, and it was a painful experience. I had no idea what I was asking for."

"I didn't understand the complexity of bringing animals to set at that time," he added. "That experience left an impression, and it just became clear when I started working on *Noah* that using animals would not work for me. It started from a practical point of view, and then it made complete sense. There was no way we could do this with the message of the film. This was a period of growth for me."

Just a few years before, when he was making *Fountain*, Aronofsky did not have as many tools to realistically depict animals as he did when he started *Noah*. The conventional way, other than dressing up humans in costumes, was to hire a trainer to force an animal to perform. But on the scale of *Noah*, that was a tough proposition. "It probably wouldn't have been physically possible to use animals with any really good results," he said. "And that's completely apart from any ethical issues related to animal treatment." The answer lay in computer generated imagery, or CGI—the ability to render lifelike animals without using, or harming, a single real animal in the process.

In a century of filmmaking, directors have sought to achieve authentic representations of animals, but struggled with the challenges of training them and the reliability of wild animals' performances. In the early days, humane advocates hoped to see improvements over the circus and vaudeville practices that were frequently so cruel to animals. However, the new medium brought its own cruelties, not relief. Animals suffered in performances, training, and housing. A blue ribbon committee convened by the *Christian Science Monitor* in 1925 recommended the elimination of scenes in which "creatures are coerced to perform unnatural and dangerous acts . . . . Bull fights, rodeos, diving horses, stampedes of herds of cattle, animals performing dressed as humans, and similar acts degrade the public taste and cause pain to the harmless creatures employed." Only on occasion did filmmakers seek to use fake animals or trick photography to achieve their effects, and serious abuse of animals continued for decades. Animal trainers stood ready to meet the demands of almost any director or studio no matter what the cost in cruelty.

The history of filmmaking included an incredible roster of animal stars, role players, extras, and other performances. From Rin Tin Tin, Old Yeller, and Lassie to Toto, Benji, and Beethoven, there have always been A-list canines in Hollywood. There have also been famous horses, including the talking Mr. Ed, and Fury, a star of the 1950s and 1960s who for a time earned $500,000 a year for his touching performances in *Black Beauty*, *Lassie*, and his own show. There have even been celebrity pigs like Babe, and chimpanzees such as Cheetah, who starred in the Tarzan movies of the early 1930s and lived on to a rather extraordinary eighty years of age. As late as 2011, fans of the Tarzan series could have visited a primate sanctuary in Florida to say hello to the very chimp who shared the screen with Maureen O'Sullivan and Johnny Weissmuller.

Used mainly in galloping and battle sequences, horses suffered in film productions, especially in westerns and war movies. More than one hundred horses perished in the making of *Ben Hur* in 1925. Scandals surrounded the production of *The Charge of the Light Brigade* (1936), in which director Michael Curtiz broke so many horses' necks and legs with the trip wire that the film's star, Errol Flynn, secretly complained to the ASPCA. Howls were also raised about *Jesse James* (1939), whose director drove a horse off a cliff to his death on set. That same year, in the classic *Stagecoach*, director John Ford used trip wire to dramatically upend the horses, with "wires attached to the horses' forelegs [that] were threaded through a ring on the cinch and secured to buried dead weights," so that "when the horse ran to the end of the wires, his forelegs were yanked out from under him." It wasn't just US filmmakers; in the 1966 Soviet-era classic *Andrei Rublev*, a horse is pushed down a flight of stairs and falls onto a spear in a ghastly scene.

While the use of domestic animals in films always raised serious concerns, the treatment of wild animals presented even more problems, since they can be entirely unpredictable. In 1903, in one of the ugliest incidents in cinema history, paying customers watched two hucksters poison, strangle, torment, and electrocute an Asian elephant named Topsy in a public display on Coney Island. Topsy, captured from the wild and brought to the United States a quarter century earlier, had faced unending abuse in a circus, and killed a man who had been taunting her by burning the tip of her trunk with a lit cigar. The public execution was a final act of punishment, but it was also one last opportunity to squeeze more revenue and publicity from a long-tormented creature. Edison Electric recorded *Electrocuting the Elephant* as a seventy-four-second black-and-white film.

Animation was an early breakthrough: even as far back as *Bambi* in 1942, it was vivid and emotionally powerful, and provided endless variety in storytelling. But animated cartoons were mainly reserved for audiences of children. Later, moviemakers used animatronics—lifelike robots, as seen in the *Jurassic Park* series. But animatronics, and even the earliest forms of CGI, frequently produced what the industry knows as the "uncanny valley," or the aversion people feel towards things that appear lifelike, but are nevertheless not quite right. The replicas looked too realistic to be animated, but too animated to be real.

Depictions of animals in film became more sympathetic in the post–World War II years, and none were better than in Disney portrayals like *Old Yeller*. Yet as decades passed, even Disney engaged in mistreatment for cinematic flare, with one of its films showing Arctic lemmings careening over a cliff and falling hundreds of feet onto rocky shores below. Lemmings don't naturally engage in mass suicidal behavior—the animals were physically thrown into the water, where they drowned, in one of the most glaring examples of nature fakery on film.

Directors typically do multiple takes to knit together a story and create a compelling illusion for viewers. Often that means hundreds of hours of filming for every hour that makes the final cut. When it comes to animals in film, the unused footage usually includes not just the injury and death that occurs on the set, but also inhumane training practices and even punishing and miserable transport. In *Every Which Way But Loose*, Clint Eastwood traveled around with an orangutan named Clyde who made audiences laugh with his performance. But animal-welfare groups charged that Clyde's trainer beat him to force him to perform. According to Jane Goodall, Clyde was trained with a can of mace and a pipe wrapped in newspaper. "Near the end of filming the sequel *Any Which Way You Can,* the orangutan was caught stealing doughnuts on the set, brought back to the training facility and beaten for twenty minutes with a three-and-a-half-foot ax handle," according to an investigation conducted by People for the Ethical Treatment of Animals (PETA). "He died soon after of a cerebral hemorrhage," the victim of a vicious bully, according to PETA.

The American Humane Association (AHA) has accepted the charge of overseeing the use of animals on set, under a contract arrangement with the Motion Picture Association of America (MPAA) since 1940. AHA personnel are on hand for some of the takes involving animals, and in return for following their directions, moviemakers can secure the "No Animals Were Harmed" tag in the film credits—a sort of good housekeeping seal of approval for the studios. In 1981, the AHA called out the makers of the film *Heaven's Gate*—a box office disaster that was also a nightmare for the animals involved. The film, set in Wyoming, produced a number of dead horses, and the director even staged live cockfights. The AHA not only denied its tag line to the movie, but organized protests at theaters.

The abuses documented during the filming of *Heaven's Gate* resulted in the AHA gaining greater access and authority on sets, but in the years since, the group has been widely criticized for lax enforcement of its animal-protection mission. Pat Derby, the late founder of the Performing Animal Welfare Society and a former trainer of animals for television, observed in 1989, "I've watched the AHA a long time, and I believe it's structured to be more of a protection agency for trainers. On every set I've ever worked on, when a rough scene was going to occur, the AHA rep was taken out for coffee." Derby and television icon and animal advocate Bob Barker have been among the most outspoken critics of the organization, arguing that the AHA's independence is compromised because the studios pay it. And with the globalization of filmmaking, and so many movies done on location in other parts of the world, the AHA simply cannot keep up with the entire industry. The AHA has no power, nor access, when it comes to the treatment of animals before they get to the set, which is often when the worst beatings and other abuses occur. A 2013 story in the *Hollywood Reporter* made it clear there were still major problems in the industry and that the AHA compact with the MPAA was not good enough to keep animals safe.

Public scrutiny and concern have dramatically reduced the abuses on movie sets, but when wild animals are used in films there are still inherent problems. In the production of *Life of Pi*, director Ang Lee filmed a tiger and a boy on a twenty-foot-long boat. Lee got plaudits for the wondrous results he achieved using CGI technology, since he couldn't after all have a boy and tiger interacting in such close quarters. But for many of the scenes he felt he needed a live tiger—and it did not go smoothly. "This one take with him just went really bad and he got lost trying to swim to the side," an AHA representative wrote in a private communication in 2011. "Damn near drowned." But the movie went to the box office with the AHA's "No Animals Were Harmed" designation. Lee got credit for his technical wizardry, and the near-death experience of his striped star was overlooked.

Darren Aronofsky, in our conversation, pointed to a recently revived historic film franchise as a harbinger of better things to come. 'I think it's incredible what they've done," he said of the latest *Planet of the Apes* movies. Directors Rupert Wyatt and Matt Reeves used CGI to depict "sentient" apes who leap, swing, fight, ride horses, and converse in *Rise of the Planet of the Apes* (2011) and *Dawn of the Planet of the Apes* (2014). As an alternative to live animal use in filmmaking, Aronofsky considers their work the new gold standard for CGI.

Behind some of that work is the husband-and-wife team of Rick Jaffa and Amanda Silver, who pitched the revival of *Planet of the Apes* and wrote the screenplays for the series for Fox Studios. Jaffa told me that he had been collecting clips on private ownership of chimpanzees and was struck that it "always ended badly" for the animals when people kept them as pets. The animals "attack a neighbor, or they bite their owner. As they get aggressive or bigger, they wind up confined in cages. They get separated from their owners. Even if they are put into a safe haven, even still, the separation is as powerful as dropping off your own child."

At the same time, Jaffa had also been taking a hard look at the issue of genetic engineering. "I had a bunch of articles spread out on the floor, and my eyes went back and forth, and a little voice came into my head—*Planet of the Apes*." He and Silver were huge fans of the original franchise, and once it gelled for them, they put the idea in front of Fox. They wanted to develop ape characters, including their primary protagonist Caesar, who would lead the ape revolution. Unlike Aronofsky's film, which cast Noah in the main role with animals in supporting roles, Jaffa and Silver would cast CGI animals as the central characters.

Jaffa and Silver noted with great admiration that director James Cameron and the visual effects company Weta Workshop had, in *Avatar*, accomplished extraordinary things by means of "performance capture." This innovation catches human expressions and movements, and then reproduces them in digital, animal-like characters. As in *Avatar*, the characters in *Rise of the Planet of the Apes* and *Dawn of the Planet of the Apes* would be entirely the creation of this process. Jaffa and Silver wrote their script knowing that Cameron and Weta had cracked the code, preparing the way for a movie that would change filmmaking forever.

The directors of *Rise* and *Dawn* used performance capture to deliver some of the most realistic CGI animal performances ever seen—a hybridized process of human acting and technological wizardry. "The actors who play the apes are wearing gray body suits, with sensors all over their bodies and markers all over their facial muscles and they are covered with dots," Jaffa explained. "All of that is fed into computers and there are cameras everywhere picking up all this stuff. There are artists and technicians who take the images of the human performances and pull it all together in the form of the on-screen images.

"We have seen countless cuts of the film where you watch actors in gray body suits, and it plays emotionally and powerfully, and you forget that they are going to be apes in about four months. When you see the apes in the movie, you are actually seeing the human performances. It's one of the most amazing things we've ever seen." They stressed that CGI technology, while a remarkable innovation, still needs a powerful human performance to give it full meaning and life. "More than a few times," they told me, "the performance from Andy Serkis—a grown man in a leotard and skull cap playing Caesar—is so powerful that people watching it are crying."

Like Jaffa and Silver, I was a big fan of the original *Planet of the Apes* with Roddy McDowall and Charlton Heston. There was something appealing and subversive in the idea of a fictional society of talking apes inverting the social order. Even then, as a child, I'd hoped that *Planet of the Apes* might stir some self-reflection in our species. The apes depicted in the movie, including the human rights advocates Cornelius and Zira, were played by humans in costume and makeup, although a few live baby chimps made cameo appearances. "At the time the makeup artist John Chambers came up with those masks—they were movable," Jaffa told me. "Chambers was an absolute genius—it was like a great leap forward, and decades later, performance capture was another." Unlike the original, the new version of *Planet of the Apes* would feature thousands of apes, each one authentic and unique. They had all of the musculature and features of real apes, and they could engage in acrobatics that human actors, even Olympic gymnasts, could never hope to replicate. "In this third act in *Dawn*, you are watching a herd of apes running across a bridge, and policemen are trying to stop them, and you are rooting for the apes because in this form, they are so emotional and compelling."

Both Jaffa and Silver told me there was just no way they'd ever be part of a movie using live animals. "*Project X* was in our minds when we were writing the *Rise*," Silver told me. "Some said there was always a stunt guy standing by with a loaded gun because they were working with real apes. That sent chills through me. It's dangerous for everybody." That 1987 movie, which exposed the irradiation of chimps in experiments by the U.S. Air Force, prompted something of a civil war within the animal welfare movement. Many advocates celebrated the idea of reaching the masses with the problems of military experiments on chimpanzees, while others said that the beating and traumatizing of chimps on set stained the movie and its message.

Project X is history now, and no chimp need suffer in a movie again. "You will get what you want out of computer-generated animals," Jaffa said to me. "You will get a better performance. You can manipulate everything—from the coat to the expressions. You get control. And eventually, it's going to be a lot cheaper."

Darren Aronofsky told me something similar. "It costs $200,000 just to do a day of shooting" with live animals, he explained. "If you have to redo scenes because the animal didn't look the right way, the costs shoot up. You avoid all of that with CGI. And to compare, $200,000 is a major visual effects shot." He emphasized that CGI characters, formed on a computer, are neither cartoonish nor wooden. "You can create a convincing illusion of the wild animals and you can have a wild animal do anything you want," he said.

Given that so many people learn about the world and are entertained through filmmaking and television, it's an important and explicit message to deliver to the masses that animals should not be sacrificed or made to suffer for storytelling purposes. With every industry that sheds cruelty to animals, it leaves other industries still relying on abuse more vulnerable and isolated, setting the stage for changes within their domains too.

And then there's the very practical matter of ending the era of exploitation of animals on the set. In the last decade, CGI has done more for animals in film—to say nothing of what's it's done to enhance the quality of cinema—than AHA has done in seventy-five years. This new technology takes live animals entirely out of the equation yet allows directors to keep animals at the center of the human story. And when you consider the immense number of animals portrayed in filmmaking throughout the world, the numbers really add up and the amount of suffering prevented is immense. You need never doubt that "no animals were harmed in making this film" when no animals were even there.

Ringling Relents

"There are endless possibilities," said Juliette Feld, executive vice president of Feld Entertainment, the parent company of Ringling Brothers and Barnum & Bailey Circus, predicting "more motorsports, daredevils, and feats of human physical capabilities" would replace its elephant acts. Like a human acrobat shot out of a cannon, Ringling's March 2015 announcement that it would phase out traveling elephant acts came with an attention-getting bang and a suddenness that none expected, startling animal advocates and other circus companies alike. For more than a century, the costumed pachyderms and their headstands and synchronized pirouettes had been the centerpiece of the self-proclaimed "greatest show on earth."

"We're always changing and we're always learning," said Kenneth Feld, who took over the family company as CEO in 1984 after the death of his father, Irvin. "I don't know any 145-year-old company that can survive without making changes. It's the best thing for our company, our family." Mr. Feld noted that Ringling's leaders through the decades had both devised and then shed acts that put human oddities on display, including a bearded lady, dwarfs and giants, and even people with disabilities and illnesses. With such degrading exhibitions, the circus has always stretched the boundaries of social acceptability, trading on shock value and base appeals to human curiosity. For decades, zoos and roadside menageries did something similar, drawing patrons by showcasing the exotic and the bizarre, at times crossing the line not only of good taste but also of decent treatment of fellow human beings. In 1906, the Bronx Zoo confined a human Congolese pygmy named Ota Benga, exhibiting him in the monkey house "each afternoon during September," apparently attempting to make an attenuated point about ethnography and the origins of humanity. Led by African American ministers, protests ensued over the dehumanizing treatment of Benga, whose presence drew throngs of visitors who gawked at the man caged with an orangutan in the monkey house and, in some cases, who harassed and menaced him when he was later permitted to roam the grounds. While the Zoo's William Hornaday defended the exhibit, as did many other opinion leaders of the day, he bowed to the threat of a lawsuit and set Ota Benga free. A decade later, the African native, still a young man, shot himself in the heart.

The writer Jack London was not the first to raise concerns about the mistreatment of animals in entertainment spectacles and films, but he greatly aided the campaign against such cruelty with his posthumously published *Michael Brother of Jerry*. In it, London decried the terrible cruelties that animal performers suffered, lamented the breaking of wild animals for such performance, and included a foreword asking children and their parents to boycott such spectacles. With his widow's permission, the American Humane Education Society, an affiliate of the Massachusetts SPCA (MSPCA), launched the Jack London Club to call for a boycott of performing animal spectacles, as London had suggested in the foreword to his book. The journal *Our Dumb Animals*, the MSPCA's monthly magazine, devoted a few pages of every issue to depictions of the sad lives that animals lived offstage, confined in small quarters, transported roughly from place to place, their most basic needs neglected. In the 1920s, sentiment ran highly enough against the mistreatment of performing animals in the circus that Ringling Bros. cancelled its large animal acts in the second half of the decade.

In the end, Ringling outlasted London's protests and legacy, but it never broke free of the moral controversy around the treatment of wild animals in acts and exhibits. Ringling and other circuses, roadside zoos, and marine amusement parks would face unyielding pressure about elephants and big cats in the three-ring circus; juvenile tigers, lions, bear cubs made available for picture-taking with the paying public at roadside zoos; or killer whales confined in concrete swimming pools at theme parks.

It was this protracted controversy over animal welfare that compelled the Feld family to yield and to retire its circus elephants to the company's Center for Elephant Conservation in Florida by May 2016 (an accelerated time frame after Ringling's original pledge of the end of 2017). "We have detected a shift in mood from some of our customers that didn't necessarily feel comfortable with elephants traveling city-to-city," Steve Payne, a company spokesperson, told the press. Mr. Feld also said it had been hard for Ringling, which had a long history of fighting critics at every turn, to operate as a traveling show with dozens of locales restricting the use of wild animals or banning the use of bull hooks to handle them. Indeed, Los Angeles and Oakland—at the urging of HSUS—had banned bull hooks on humane grounds just weeks before Ringling's elephant retirement announcement. "I can't fight city hall," said Mr. Feld.

With elephants at the very center of the company brand of Ringling Bros., the news was something like the fall of the Berlin Wall for the animal-protection movement. Relative to factory farming or animal testing, circuses use few animals—perhaps tens of thousands, if you add them up all over the world, as compared to millions and billions in the biggest of animal-use industries. But the symbolism was incomparable because you'd be hard pressed to find a kid who hasn't attended a circus; it's been something of a cultural rite of passage for a child to go to Ringling, which each year spent millions to draw in kids by advertising elephants as the featured performers. Precisely because it was Ringling, the decision to stop using elephants had—like the animals themselves—special force and weight. No other circus had Ringling's resources or resolve in defending the animal-based circus as a business model—sparing no expense in enlisting lawyers and lobbyists to fight reform at every turn. With Ringling finally closing the curtain on this act, the politics of using elephants in circuses—and perhaps even other wild animals, too—was upended. It was a signal moment in the progress of the humane economy, demonstrating that even the most familiar and supposedly benign forms of cruelty can eventually give way to change for the better.

It had not been a smooth road, however. Over the years, Ringling's Ken Feld earned a reputation for ruthlessness and cunning in dealing with his adversaries. In the early 1990s, Feld hired Clair George, a former deputy director of operations for the Central Intelligence Agency, to lead a series of clandestine operations to gather intelligence on his foes and to distract or disrupt them. One target was Janice Pottker, a freelancer who wrote a 10,000-word piece on Feld Entertainment for the Washington, D.C., area business magazine *Regardie's*. When Pottker signaled that she wanted to dig in deeper and write a book about the Feld family, Feld unleashed Clair George. George in turn enlisted an operative, Robert Eringer, to befriend Pottker and disrupt her project, according to thousands of pages of documents unsealed in a later legal dispute. This intrigue directed against a largely unknown journalist was an eight-year gambit for Feld. It was utterly disproportionate and overreaching—Pottker estimates that Feld spent $3 million on the surveillance and disruption effort.

George, who had been drummed out of the CIA after his involvement in the Iran-Contra affair, also allegedly oversaw the infiltration by Feld operatives of at least two animal-protection groups critical of the circus—People for the Ethical Treatment of Animals (PETA) and the Performing Animal Welfare Society (PAWS). Feld got into protracted legal fights with both of those groups, ultimately paying PAWS an undisclosed financial settlement and giving the group several Ringling elephants, who were retired to the organization's California sanctuary.

Subsequently, Feld became embroiled in a legal bout with another set of animal welfare groups, which sued Ringling for allegedly violating the Endangered Species Act by routinely chaining, beating, and abusing Asian elephants. That case, originally led by the ASPCA, the Animal Welfare Institute, and The Fund for Animals, lasted all of fourteen years, and in the end turned out well for Feld. US District Judge Emmet Sullivan denied the animal-protection groups standing to sue and never ruled on the merits of the original claim against the circus. As was customary for Feld when he felt threatened, he went on the offense, filing a civil RICO case. The legal saga—both the original case filed by animal organizations and then the civil RICO claim by Feld—was settled just 10 months before Ringling's announcement to retire the elephants. The animal groups (including HSUS, which was brought into the case after it combined operations with an original plaintiff, The Fund for Animals) ended up paying Feld's legal fees but did not admit to any of Feld's far-fetched claims in the RICO suit; in fact, the litigation only exposed some further damning information about the treatment of animals by Ringling.

So how could a guy willing to go to these lengths—spending millions on public relations, infiltration efforts, and endless litigation—pivot so sharply, especially after his most recent lawsuit went his way? It may have been fatigue from decades of battle over animal-welfare issues against both animal groups and government regulators. In November 2011, the US Department of Agriculture fined Ringling $270,000 to settle numerous alleged violations of the Animal Welfare Act, including forcing an injured elephant to perform, allowing animal escapes, and losing control of an elephant inside a crowded arena. The fine against Ringling—though inconsequential against Feld Entertainment's hundreds of millions in revenue—was nonetheless the largest animal-welfare penalty ever meted out in the agency's history and a black mark against the company.

But more likely Ken Feld was looking ahead—seeing no end to

the political fight, with more local ordinances expected to germinate throughout the country. I had learned from a friend that Feld Entertainment had hired a political consulting organization, Purple Strategies, and its work revealed the unavoidable truth: years of campaigns on behalf of captive and performing wild animals by animal groups had slowly opened the public's eyes to the inherent problems of using elephants in traveling acts. With investigators producing behind-the-scenes footage of elephants being badly abused, USDA reports of problems, and protestors massing outside of circus arenas and making the case for reform, it was difficult for a discerning person not to have doubts about the ethics of using elephants in the circus. Purple Strategies made it clear that Ringling might be able to fend off a cascade of local regulations, but only by resorting to a costly public relations campaign. For a company that had a broader portfolio of subsidiaries—from monster truck performances to Stars on Ice—what was the point? Why not invest more in these other forms of noncontroversial entertainment instead of waging a losing, protracted battle over animal acts? If it was the principle of the thing, what principle could conceivably justify the torment and degradation of noble creatures like elephants? Feld had money but not the high ground by anyone's measure. Ringling had succeeded in transforming itself many times—from tent shows and the display of human oddities to the blend of human and animal performers in its contemporary three-ring acts. And Feld Entertainment had already diversified its entertainment portfolio, which now brought in $1 billion every year, with the circus accounting for a smaller and smaller share of that total.

Through the years, Feld and animal advocates did agree on one thing: Asian elephants are extraordinary in so many ways, with their gargantuan size, long and dexterous trunks (operated by forty thousand muscles), and ability to fascinate people with their very presence. They had, for decades, been the not-so-secret sauce in Ringling's formula for success, as well as the main provocation to animal advocates. Ringling had long traded on the public's fascination and love for pachyderms, but that made the business vulnerable, since there was a backstory of exploitation that had the potential to alienate the fans who came, in large part, because of their affection for elephants. The circus, like other enterprises that use and harm animals, is involved in a sort of never-ending cover-up—using all the right rhetoric in talking about its devotion to the animals and its high-caliber animal care, even as its handlers beat the animals with bull hooks and kept them in chains.

As with wild animals in the movies, it isn't so much the handling of the animals during a shoot or a live performance, but what happens before and afterwards, in the shadows where nobody's around to watch. It's in the holding tents and the shuttered boxcars where the animals endure constant privation and flashes of violence. Such mistreatment may be motivated by the handlers' exasperation over the animals not doing as they're told, or by the pressures of having to keep them in line week after week. After all, they're dealing with animals whose emotional states vacillate between fear and frustration. Dealing with unhappy eight-thousand-pound animals is never easy, and these elephants are denied life with their family members, subjected to never-ending travel and confinement in deficient housing, and chained for hours, standing on nothing but concrete or compacted dirt. They may be on the road forty-eight weeks a year, including cold weather months in places where the temperature might dip below zero—close to one hundred degrees short of what the thermometer might read in the Asian elephant's natural habitat in India or Indonesia. They almost never get to stand on a patch of grass, pull leaves off a tree, or feel the caressing touch of a mother or baby. It's endlessly boring except for the training sessions, where they endure battery, and the performances, where they work under the threat of violence and are forced to do things completely alien to their nature.

The problem starts with the need for the circus to control the elephants and the asymmetry in size between elephants and their handlers. The elephants are generally about thirty times heavier than an adult male trainer. They're aware of their advantage in size and power, which is why in the wild a healthy adult elephant does not cower in the presence of a water buffalo, a rhino, or even a tiger. In fact, with the breakdown of elephant families due to widespread poaching, some young African elephant bulls will now attack rhinos, who themselves may weigh three tons. If they could topple and kill a creature of that size and strength, imagine how easily they could turn a person into a ragdoll. Even the king of beasts knows better than to fool around with a full-grown elephant. So a baboon, a chimp, or any other primates in the wild (the closest animals to a human primate) are hardly a source of concern to an elephant. In a captive setting, however, in the presence of a human trainer or handler, elephants submit, because there's a history of breaking them through violent beatings and being kept in line through frequent reminders of the pain that can be inflicted if necessary. The handlers deliver that pain by wielding a bull hook—a hybrid of a wooden bat and a sharpened fire poker—and jabbing, prodding, and striking the elephants on their sensitive skin with its sharp end. At Ringling and other circuses with elephants, there's always a guy milling around with a bull hook clasped in his hand—in training, during the procession from the holding facility to the arena, and even in the performance ring with thousands of spectators watching.

While a bull hook is certainly intimidating, it's not always enough to control an elephant. Janet, a female Asian elephant with the now defunct Great American Circus, went on a rampage in 1992 in Palm Bay, Florida, with five children on her back. A trainer sidled another elephant up next to her, and pulled the kids off her back before she could hurt them. Police officer Blayne Doyle shot her forty-seven times before she died. Doyle, shaken by the horror of what he witnessed that day, has been active ever since as an advocate to end the use of elephants in circus acts.

Perhaps the most horrifying incident occurred two years later. Tyke, a female African elephant—who, like other elephants, had been ruthlessly beaten with bull hooks—had been rented out by the Hawthorn Corporation to the fly-by-night operation Circus International for a performance in Honolulu. Shortly before she was to perform, Tyke attacked a circus groom, violently kicking him around until her trainer tried to intervene. In front of a horrified crowd packed with children, Tyke knocked the trainer to the ground, bent down, and in several swift moves crushed him with her head and knees. With the handler sprawled out and bleeding and no one else around to stop the eight-thousand-pound elephant, she charged past frightened spectators, through an arena exit, and ran out of control and terrified into the streets of Honolulu, kicking and nearly killing a parking attendant before she was distracted. She ran free for nearly thirty minutes, with people fleeing in all directions as police on foot and in cars gave chase. Police opened fire on Tyke, but without weapons that could readily bring her down, it took nearly one hundred shots, most from nine-millimeter handguns, with the poor creature riddled with wounds before she died. Tyke had been at the center of two other uncontrolled rampages just a year prior. In one of those incidents, at a state fair in Minot, North Dakota, she had kicked and trampled an elephant show worker, breaking three of his ribs, and then led sheriff's deputies and circus workers on a twenty-five-minute chase as she ran through a campground and into a maintenance building. Despite this history of unpredictable breakouts, Hawthorn was content to shuttle her around the country for performances in arenas crowded with thousands of people.

Tyke and Janet had had enough, and they lashed out in fury. In most instances, though, fear fosters obedience, and the elephants submit to their daily dose of punishment, like automatons faithfully performing a set of silly tricks in the arena. Their forbearance is remarkable given what they endure. "They are required to perform behaviors never seen in nature," says conservationist and researcher Cynthia Moss, who has spent forty-five years studying elephants in the wild. "In short they are treated as commodities, as objects to provide entertainment for humans. The circus experience has nothing to do with the reality of elephant life and behavior."

Even longtime Ringling employees concede that the circus frequently uses bull hooks and chains. Robert Ridley, who worked for Ringling for more than forty years, testified that he saw "puncture wounds" on the elephants caused by bull hooks three or four times a month. He also testified he saw "hook boils"—infections caused by bull hook wounds—on average twice a week. One Ringling memorandum reported that an elephant had sustained twenty-two puncture wounds, presumably left by a bull hook. A separate internal memorandum from Ringling's own animal behaviorist reported that an elephant was bleeding "all over the arena floor" because she was hit with a bull hook several times during a show. According to Ringling Bros. emails, "lacerations" were observed on elephants from the "hooks" used on them during their morning baths. Just brandishing the bull hook provides a constant reminder to elephants of the painful punishment that can be meted out against them at the whim of their handlers. Kenneth Feld himself testified under oath that all the company's trainers used bull hooks.

As more is known about the intelligence of elephants, coercive training has fallen from favor in most quarters. The Association of Zoos and Aquariums (AZA), the trade association for about 220 accredited zoos and aquariums, adopted new elephant care standards in 2014 that prohibit bull hooks and direct contact and instead endorse "protected contact" with elephants. "Protected contact," which began to gain favor about twenty years ago, uses treats and praise to induce elephants to behave as needed in order to care for them in captivity. The Detroit Zoo went further, closing its elephant exhibit in 2002 after its forward-thinking director, Ron Kagan, concluded that he and his staff could not provide the animals with the complex enrichment they require or a suitable climate. He transferred the zoo's two elephants to a sanctuary, where they can now roam hundreds of acres in a warmer climate. Mind you, the Detroit Zoo did not keep the elephants on chains, use bull hooks, or transport the animals around for shows—but even in the absence of such practices, Kagan and his staff colleagues still felt they were unable to give the animals a decent life. Others have reached the same conclusion, and by 2015 officials at zoos in Chicago, New York, and San Francisco had retired elephants with no plans to replace them, and zoos in a host of other cities were under pressure to do the same thing.

In almost every respect, circus elephants have it much worse than those living in zoos. While the circus clowns and tightrope walkers traveling with the circus rest in beds on trains or in mobile homes, elephants stand chained in boxcars as they go from city to city. Ringling's performing elephants might go to as many as 115 cities a year. Captive elephants do not live with family members as they do in the wild. And while elephants in the wild may travel forty miles a day, knock down trees and pull two hundred pounds of grass from the ground or bushels of leaves from trees, and bathe and play in mud and watering holes, Ringling's elephants live in chains almost constantly. Ringling's own "transportation orders"—the contracts the circus enters into with railroad companies to use their tracks to carry live cargo—documented that, on average, the elephants are chained by two legs on hard surfaces, on narrow dark railroad cars for twenty-six consecutive hours when in transit—about forty-eight weeks each year—and that they are often chained for sixty to seventy hours at a time, or longer.

Elephants are one of the few species—orcas are another—with shorter life spans in captivity than in the wild, where animals are exposed to predators, droughts, and other threats. In the past ten years tuberculosis has been diagnosed in as many as nineteen of Ringling's fifty or so elephants, according to a PETA report based on USDA documents. Tuberculosis is a serious and potentially life-threatening disease that is unheard of in wild elephants but common among captive elephants, perhaps because they are chained in close proximity for hours, and stressed by the rigors of travel, extremely cold temperatures, and the fear of their handlers.

The Ringling announcement was a jolt that lifted the moral confusion and conflict that had surrounded the issue for years. A once largely silent news media now began to echo and recapitulate the longstanding criticisms that animal advocates had been making. In the days after the announcement, the *Boston Globe* called the move "terrific news for a majestic species of animal" and "a triumph of public awareness," while the *New York Times* declared, "these magnificent creatures deserve better than being dolled up and sent on the road to do stunts." The *Dallas Morning News* wrote, " 'The Greatest Show on Earth' is soon to be even greater because Ringling Bros. and Barnum & Bailey Circus says it will end elephant performances." Ringling's hometown newspaper, the *Sarasota Herald Tribune*, observed that Ringling's decision "represents the evolution from the old circus to the new." The *Cincinnati Enquirer* editorial board peeked in on Ringling spokesman Stephen Payne while he was talking about the company's treatment of animals with a group of elementary school students, who weren't buying the notion that Ringling's elephants are happy. "If you are going to do more shows, do shows with just people. No circuses with animals," said eight-year-old Sohini Mallick at the Fairview-Clifton German Language School. "If you think that I will be entertained with animals, I actually don't. I would be entertained with only people."

Just a month after Ringling's announcement, another domino fell. Bill Cunningham, owner of the George Carden International Circus, said it too would phase out its elephant acts. Cunningham explained that having elephants perform tricks "doesn't appeal to our higher selves and I think we as a society have evolved in too many other areas for that practice to continue unchanged." He added, "There are so many different types of entertainment available that could be included in [the] circus, and advanced technology to present acts in compelling ways, so why continue with a genre—performing elephants—that seems outdated and alienates potential fans?"

In the years before Ringling and George Carden saw the light, a number of circuses gave up their animal acts, including the Big Apple Circus (2000), Circus Flora (2000), and Circus Vargas (2010). More than two dozen nations had banned the use of elephants for entertainment, including Austria, Bolivia, Cyprus, and Peru. In Australia and Canada, by 2014, reformers had secured bans in twenty-eight municipalities. It was a global movement against wild animal acts, and it was gaining ground.

Feld did not have to wait long for additional evidence of the political and practical opposition that Ringling would have faced had it kept its elephant acts. In May 2015, just months after Ringling's announcement, Governor David Ige of Hawaii pledged to stop issuing permits for any wild animal acts to visit the island. Months later, California lawmakers banned bull hooks by overwhelming margins; Governor Jerry Brown vetoed that bill, but only because of his quarrel with a proposed enforcement mechanism and not because he had any sympathy for the use of bull hooks. And other nations took action to ban wild animal acts, including Mexico and the Netherlands, which had sixteen circuses with wild animal acts.

But competition in the circus world also explains Ringling's turnaround. Some circuses, most famously Cirque du Soleil, have never had elephants or other animals. Since Guy Laliberté and Gilles Ste.-Croix launched Cirque du Soleil in 1984 as a "group of stilt-walkers and street performers" out of Quebec City, it has featured amazing acrobatics in meticulously choreographed, death-defying performances. By 2012, it was generating $1 billion in revenue—about as much as all of Feld's entertainment operations, including the Ringling circus. Las Vegas was once known for its animal acts—including *Siegfried and*

Roy, until a performing tiger grabbed magician Roy Horn by the neck in 2003 and just about killed him, ending the show permanently. But today there's hardly a wild animal act to be found in Vegas, while this city known for its live performances and entertainment now has eight Cirque du Soleil productions. Guy Laliberté, who sold a majority stake in the company for $1.5 billion in April 2015, has nineteen productions globally and "sells eleven million tickets worldwide annually, more than all Broadway shows combined." The private equity group that purchased such a large share of Cirque du Soleil from Laliberté has plans to expand in China and take advantage of the rapidly growing leisure and entertainment market there. Laliberté and Ste.-Croix started their company the same year Ken Feld took over Ringling from his father, and the social reaction to the two different business models is striking. It's impossible to imagine a major private equity group buying Ringling or some other circus because of its animal acts. It's the company without animal acts in which the smart money people are investing their billions.

With pressure from animal advocates and the changing values and tastes of the public, we may soon see the curtain close on all wild animal acts, which look archaic compared to other attractions. Compare a Cirque du Soleil performance with an elephant show at Ringling, and it's like a proxy fight between the new and the old—one an innovative display of human acrobatics and choreography and the other a vestige of an earlier era when we entertained people with freak shows and suffering wild animals. We are at a watershed moment, and there is no going back now that we realize, as columnist Charles Krauthammer wrote, "festooning these magnificent creatures with comically gaudy costumes and parading them about, often shackled, is a reproach to both their nobility and our humanity." As the number of animal acts shrinks, the remaining performances with animals will look all the more out-of-place and inhumane. Circuses that continue to use elephants or rent them out are managing an aging herd, given that Asian elephants are now endangered and hard to import from the wild or breed in captivity. Nobody in the circus industry is thinking about where the next generation of elephants will come from, since everybody expects the transition away from animal acts to be completed soon. It's just a matter of time before the question becomes where these circuses' old elephants—with their sore and broken feet, their severely scarred skin, and their broken spirits—will be retired. At Cirque du Soleil and other animal-free circuses, customers don't have to walk past or through picket lines, or wonder whether the animal performers were kept on chains for eighteen or twenty-two hours a day. The company need not worry about social media campaigns, shareholder resolutions, or local ordinances to prevent it from coming to a city. Its performers choose their vocation and are compensated for it, and the patrons have no festering qualms or complaints or concerns about animal cruelty. Consumers now have options, and they are already making their feelings felt and reshaping the marketplace. Dazzling displays absent animal cruelty is the new economic reality, and who will miss the old realities of boxcars and bull hooks?

See World in a New Way

"I am just a consummate question asker," Gabriela Cowperthwaite told me, explaining why she chose the topic of orcas at SeaWorld for her movie *Blackfish*. "I wanted to know why a top-level SeaWorld trainer got killed." That question, and the film that came in its wake, would roil the waters at SeaWorld, whose theme parks in Orlando, San Antonio, San Diego, and other cities are among the most visited in the nation, collectively drawing twenty-four million patrons a year (including eleven million who see the orca shows) and generating billions in revenue. *Blackfish* inspired a national discussion about whether that kind of business—which keeps the world's biggest and most powerful predators as captives, trains them to do trivial tricks, and trades on our fascination by putting them on display often just a few feet away from us—can ever be justified.

Most documentarians measure their audiences in the thousands, but among the small number of breakout documentaries, some of the most successful and widely viewed have exposed the mistreatment of animals. In 2009 alone, Robert Kenner's indictment of factory farming, *Food Inc.* (2009), and Louie Psihoyos' Academy Award–winning thriller about the slaughter and capture of dolphins, *The Cove*, were critical successes and found large public audiences. But with *Blackfish*, Cowperthwaite drew twenty-one million television viewers in one week on CNN alone—an unheard of audience for a low-budget documentary on a serious topic. "We were sold a bill of goods by SeaWorld," Cowperthwaite explained. "Sure, adults can take it, but they were selling it to our children. The company says it gives our kids educational and inspirational experiences, but it's exactly the opposite. It feels like a fraud, and when they learn the real story, people said we want our money back."

As with Ringling Bros.' abuse of elephants, or past filmmakers' beating of chimps or orangutans or tripping of horses, there was a painful backstory with SeaWorld's treatment of Blackfish, the name that first peoples of coastal North America gave to orcas. Upon its theatrical release, *Blackfish* was hailed by reviewers and generated a loud buzz. It gained still more momentum when SeaWorld issued an open letter urging movie critics to ignore the film's "inaccurate" storyline. The breakthrough, though, came when CNN bought the rights and decided to broadcast it. After CNN executives realized they'd made the right bet on the film, the network doubled down—and then some. It decided to run serial rebroadcasts, and whether you tuned in because of keen interest or mild curiosity, or got hooked while channel surfing, it was hard, at least for a time, to avoid *Blackfish* and its biting indictment of SeaWorld.

SeaWorld's stock careened downward, and Wall Street analysts took note, stoking still more interest in the film. With SeaWorld's share price down 40 percent a year after the film's premiere, its largest shareholder, the Blackstone Group, realized the damage would be as indelible as the black markings on the whales. In a public relations contest of Blackstone versus *Blackfish*, the private equity giant didn't like the odds. Having just acquired SeaWorld in 2009, it now sold off a big stake. In an even more significant no-confidence vote, SeaWorld's own CEO quietly sold $3 million of his SeaWorld shares. Rating agencies downgraded the stock. Southwest Airlines, a long-time sponsor, decided to terminate its partnership with SeaWorld. A concert series planned for Shamu Stadium in early 2014 went on, but without its big-name acts. Barenaked Ladies, the Canadian rock band, was the first to cancel, reacting to a petition posted on Change.org. Willie Nelson was next, explaining, "I don't agree with the way they treat their animals." Then cancellations came from Trace Adkins, Trisha Yearwood, Cheap Trick, Heart, Pat Benatar, Martina McBride, 38 Special, and, appropriately, the Beach Boys. A choreographed celebration of SeaWorld turned into a stampede away from it.

Within a year of the film's release, attendance at SeaWorld was down by a million, with schools switching their class trips to Disneyland and other destinations instead. (My ten-year-old niece and eight-year-old nephew called me with anger in their voices after they watched the film, asking what they could do to stop SeaWorld from "hurting whales.") In April 2015, Mattel yanked its SeaWorld Barbie from toy store shelves. About the same time, SeaWorld gave notice to more than three hundred employees, including its CEO. While delivering a speech at the University of Oklahoma business school, I was astonished to learn that just about all the students had, on their own, watched *Blackfish*, either on CNN or Netflix—and to a person, they had an unforgiving view of SeaWorld's practices.

The humane economy is more than new ideas and new technologies; it is also timeless moral values reasserting themselves and gaining force with the spread of knowledge. And when industries rely on keeping their customers in the dark to do what they do—and that's a standard operating procedure for many controversial animal-use enterprises—they are on the wrong side of the information age. *Blackfish* showed just how quickly and powerfully the spread of information can upend an unethical enterprise. No matter how big, no animal-use company with questionable practices is safe from one well-made documentary that tells the story behind it.

In the years prior to *Blackfish*, there had been plenty of controversy surrounding SeaWorld, but the company had been able to fend off the critics and keep growing, with revenues up from $100 million in 1990 to $1.4 billion in 2012. During the 1990s, however, it became a consensus position among animal welfare groups that there was something terribly wrong with keeping cetaceans in captivity—driven by the psychological and behavioral effects stemming from long-term captivity for these remarkably intelligent and sociable creatures. In 1993, the HSUS hired Dr. Naomi Rose, a marine mammal scientist who had studied orcas in the Pacific Northwest, and she proceeded to lead a national campaign to expose the inherent problems of keeping orcas in small swimming pools, living apart from their families.

Dr. Rose, Patricia Forkan, Paul Irwin, and other HSUS personnel played a big role in the effort to rescue and rehabilitate Keiko, an orca captured in 1979 in the waters around Iceland, taken from his family when he was just two years old, and confined to a series of marine parks where he was forced to perform. The 1993 feature film *Free Willy*, a blockbuster production by Warner Bros., told of a boy taken with the idea of freeing a captive orca from a dingy marine park and returning him to his family in the ocean. *Free Willy*, which grossed more than $150 million, relied in part on footage taken of Keiko in his real-life environment at an amusement park in Mexico. The public sympathy stirred by the movie prompted a movement to get the real whale out of that facility and back with his pod in the ocean. Philanthropist and newspaper publisher Wendy McCaw pitched in with one of the biggest investments ever in animal protection—joining Richard and Lauren Shuler Donner, the producers of *Free Willy,* and thousands of school children in putting up $7 million to buy Keiko from the Mexican park and then helping to build a facility in Newport, Oregon (a much improved holding facility until he was ready for release into a sea pen or the open ocean). During his time in Oregon, with professional caretakers, Keiko gained over a ton of weight in a far more hospitable living space.

After McCaw's heroic efforts and investment, and the schoolchildren's campaign across the nation, there was a dramatic airlift of Keiko from Oregon to Iceland. Millions of people worldwide followed his release into a sea pen. There he could feed on fish and do what whales do, having the run of a section of the ocean bound by the shore and a surface-to-ocean-bottom net that introduced him to his natural habitat. Later on, he was set entirely free and actually swam all the way to Norway. At age twenty-six, he died from pneumonia. Advocates of his transfer to Iceland hailed it as a success, evidenced by the fact that Keiko had been given a fuller, richer, and more natural taste of life as a free orca.

After years of cultural education, campaigns, and debate over the fate of the orcas, *Blackfish* delivered something of a closing argument. Cowperthwaite's documentary was a clarion call that the era of whale confinement must end, with marine mammal scientists translating the latest information about the complex lives of whales in the wild and former SeaWorld trainers and hunters lamenting their own roles in the exploitation of these animals.

For Cowperthwaite, who had planned a career in political science before falling in love with film, it was the violent death of a forty-year-old orca trainer in late February 2010 that led her to SeaWorld. "I had taken my kids to SeaWorld years before, and I wanted to understand why this terrible incident happened and why people were swimming with apex predators," she added. "I didn't think there was any funny business. I entered the project with a question, not an argument."

The incident occurred on an unseasonably chilly and overcast day, with attendance smaller than usual at Shamu Stadium in Orlando. Tilikum had performed a number of head shakes and dorsal fin maneuvers to tempered applause, and his trainer Dawn Brancheau had, in turn, rewarded him with fistfuls of frozen herring. SeaWorld executives allowed trainers to ride the backs of orcas, cling to their bellies as they went under water, and launch from their rostrums ("beaks") and sail thirty feet into the air—a rather astonishing display of faith given the orcas' size and predatory ways. But there was a different set of rules for trainers in dealing with Tilikum. He was the largest of SeaWorld's orcas, at six tons, and his rap sheet was scary enough that the park kept it shrouded in secrecy. He'd killed a female trainer, twenty-year-old Keltie Byrne, in 1991 at a makeshift marine park in Victoria, British Columbia. After the drowning of Byrne, operators shuttered the place and put the facility's three killer whales on the open market. That's when SeaWorld bought Tilikum and flew him to Orlando, where he was placed with a larger group of whales, who, Cowperthwaite told me, beat him up mercilessly. In 1999, Tilikum apparently killed twenty-seven-year-old Daniel Dukes, described by the company as a "drifter," who got into the park after hours and into the whale's nighttime pool. By morning, Tilikum had the dead man, absent his genitals, splayed on his broad and long backside as Tilikum made tightly wound circles in the small pool where he slept at night.

Years later, as a "Dine with Shamu" performance was winding down—patrons were served lunch while they took in the show—Brancheau got down on her chest in a prone position on a rocky ledge accessible to the whale. Then, Tilikum, who had been mildly scolded by Brancheau for not picking up or following instructions earlier, bit down on Brancheau's left forearm, did a belly roll, and pulled her into the water. Expressions in the small crowd turned from amusement to concern to dread; spectators realized the show had gone off script and an attack was in progress. Tilikum shook Brancheau at the surface, thrashing her as she vainly struggled to escape. With the woman in his grip, he dove to the bottom of the thirty-foot-deep pool, letting her go occasionally and ramming her, before again clamping down on her and preventing an escape. The assault, which ranged from the bottom of the pool to the water's surface, continued for an interminably long period of more than thirty minutes. Trainers and other personnel had been slapping the water, throwing food to Tilikum, and otherwise trying to distract or redirect him, but to no avail. Eventually, but in all likelihood long after Brancheau was dead, they coaxed Tilikum into a small adjoining pool, though he still had hold of the trainer's body. They put a net over him, and pried free Brancheau's battered and limp body, or what remained of it. Her scalp and one arm had been torn off. The autopsy report, a month later, concluded the death had come as a result of "drowning and traumatic injuries," and noted that she'd sustained blunt force trauma to her head, neck, and torso; a severed spinal cord; multiple fractures; and lacerations across her body.

While SeaWorld's official position was that Brancheau followed all safety measures, an expert it hired attempted to frame the tragedy as Brancheau's breach of protocol. In effect, he was blaming the victim to deflect attention from SeaWorld's own responsibility for allowing trainers to swim with intelligent ocean predators able to kill even blue whales twenty times their size. "The only thing that led to this event was a mistake made by Ms. Brancheau in allowing her long hair to float out into an area that Tilikum could grab in his curiosity," wrote the expert witness in a submission to the Occupational Safety and Health Review Commission, which opened an investigation into Brancheau's death and worker safety issues.

OSHA came to a different conclusion—finding that the company, not the helpless Brancheau, was at fault. In August 2010, the federal agency issued a "willful" citation against SeaWorld, meaning the agency considered SeaWorld to have shown "plain indifference" to its employees' safety. SeaWorld contested the citation. After a nine-day proceeding, administrative law judge Ken Welsch upheld the citation but downgraded it to "serious." Welsch concluded that SeaWorld believed it was doing enough to protect the safety of its trainers, but that it was badly mistaken and with grievous consequences. The judge determined that Tilikum had grabbed Brancheau by the arm rather than the ponytail, that the trainer had followed all the rules, and that the SeaWorld "expert" who testified had neither talked to eyewitnesses nor read the autopsy report. Yet as late as 2014, the chief executive of Blackstone still appeared to be misinformed, telling CNBC that the deceased trainer "violated all the safety rules that we had." To its credit, Blackstone apologetically recanted the statement the very next day, admitting it did not "accurately reflect the facts."

SeaWorld also contended that Tilikum's attack was a one-off event that it could not have anticipated. But that argument didn't track either, with the judge finding that SeaWorld knew from a series of accidents that its trainers were at risk when they were in the pool with killer whales, and had decided to continue the practice anyway. Welsch upheld OSHA's core claim that SeaWorld must abate the hazard posed to trainers by keeping them out of the water when orcas are performing—precisely the outcome that SeaWorld wanted to avoid, because it viewed the trainer–orca interactions as the sizzle behind its shows.

SeaWorld responded to the ruling by "lawyering up"—hiring Eugene Scalia, the son of a Supreme Court justice—to appeal the decision to the US Court of Appeals for the District of Columbia Circuit. But even with the company pulling out all the stops, it couldn't turn around the decision. The appellate court in April 2014 found that the administrative law judge got it right: SeaWorld knew the risks to its trainers and had a duty to protect them by keeping them out of the pool. In the federal proceedings, that would be the final word.

In May 2015, in response to a worker safety complaint, the California Division of Occupational Safety (CDOS) fined SeaWorld $25,770 for violations at its San Diego facility. CDOS cited SeaWorld for its failure to protect employees and supervisors who "rode on the killer whales and swam with killer whales in the medical pool" and "who were present on the slide outs with killer whales in various pools." In its report, the State of California criticized SeaWorld for requiring trainers to sign confidentiality agreements that discouraged them from raising safety concerns for "fear of reprisal"—the sort of measure that factory farms use to forbid workers from speaking out against abusive practices they see and wish to report.

In the wild, orcas live in matriarchal societies with large extended families. They may travel one hundred miles a day in these distinct family groups. Scientists believe that the pods maintain their own communication patterns and cultures, with tailored feeding techniques and prey depending on where they live. Some pods hunt whales, others pursue seals, while still others eat salmon and other fish. But in captivity, there is no stimulation through hunting behavior, little of the companionship that comes through family living, and limited exercise and other forms of physical or emotional stimulation. While the orcas do brief performances—some of the only stimulation they receive—they are kept for many hours of the day and through the night in small, shallow pools not much larger than their bodies.

At a certain level, SeaWorld's sophistry is extraordinary. At its shows, the company excites the crowd by hyping the physical size of the orcas and their status as the world's most powerful predators. We as a society commonly refer to the animals as "killer whales," which has become a synonym for orca. We comfortably describe and designate them as lethal, yet SeaWorld staff members were getting into pools with them for seven shows a day at each of three facilities. It's one thing to get in a pool with some species of baleen whales, which are also enormous but at least have no teeth and feed principally on small crustaceans. But with animals who travel in packs and are built to kill elephant seals and walruses and fin and humpback whales, and have been known even to kill great white sharks, it goes beyond reckless to put people in this kind of jeopardy. Consider, too, their highly developed brains, and all that comes with that—calculating actions, deception, grievances, and mood swings driven by the frustration of boredom, extreme confinement, and deprivation. Then there's the psychological trauma of being wrested from their families or forced to live in pools that in no way approximate their natural habitat. The risks of dealing with a ten-thousand-pound predator with speed, power, and sharp teeth are compounded.

In listening to interviews of trainers who survived attacks, they never seem to blame the whales. They love them and attribute the incidents to the normal ups and downs that animals feel and the complex social relationships within the group of whales and with the trainers themselves. The best of them say they can anticipate and avoid these problems at times. But workplace safety should not depend on that kind of intuition and luck. The labor issues are as black and white as the orcas, and treating the pools as mixing bowls for humans and killer whales is an invitation to injury and death. Exactly for that reason, OSHA concluded that the business of swimming with killer whales, especially after's Brancheau's tragic death, must end.

According to Cowperthwaite, there have been as many as seventy safety incidents at SeaWorld, with orcas slamming into trainers, or in some cases, grabbing them and pulling them beneath the surface, as Tilikum had done with Dawn Brancheau. In a scene memorialized on video, and included in *Blackfish*, a dominant female killer whale named Kasatka grabbed the foot of trainer Ken Peters in 2006, dragging him to the bottom of the thirty-foot pool multiple times—the second major incident he had with the whale. With extraordinary calm, and using all of his diving and breathing skills, Peters avoided drowning over the eight-minute ordeal, and then the whale released her grip. Seizing the opportunity, Peters slowly and calmly used hand-over-hand maneuvers to pull himself toward Kasatka's tail and away from her mouth and then swam for his life. He dragged himself over a net in the pool and scampered onto a slide-out, with the killer whale in pursuit. He then hauled himself out of the pool, lucky to be alive and only sporting some broken bones and torn ligaments in his foot.

In a separate incident, trainer John Sillick felt the full force of a whale landing on him—fellow trainers observed that only "the wet suit kept him together." Another trainer, Tamarie Tollison, was pulled into the water by a six-thousand-pound orca named Orkid, and bitten severely by a killer whale named Splash who was also in the tank. A quick-thinking trainer ordered the chain off the enclosure that held Kasatka, the dominant female of the group. The mere threat of Kasatka coming into the water and raking them with perfectly formed teeth caused Orkid and Splash to abandon their attack, and trainers quickly pulled Tollison out of the water. She had multiple wounds, including a compound fracture of her arm, but she did escape.

Indeed, whale-on-whale violence is one of the most significant welfare issues at SeaWorld. "They are fighting with each other pretty constantly," Cowperthwaite told me. "They are all stuck in this tank and you think they have each other and that gives you solace. But you find out that so much of their stress comes from battling one another because they don't have space. They vie for dominance constantly, and they are not from the same pods, they don't speak the same language." The females are dominant, and Tilikum, despite his size, was beaten up by female orcas on a regular basis. Precisely because of his mass and the small physical area he had to operate within, the trainers observed, "he was not as mobile, and he couldn't get away."

In captivity, Tilikum and other males have a very distinct and symbolic way of signaling to the world that something is wrong—their dorsal fin, which should stand erect, droops and falls to one side. While dorsal fin collapse occurs in less than 1 percent of male orcas in the wild, it occurs in 100 percent of captive males. Tilikum had another reason to be unhappy—SeaWorld isolated him after he was roughed up by the female whales. Tilikum, especially after the Brancheau incident, was kept apart from people and other whales—"kept in the back and brought out at the end [of the show] as the big splash." Neither option—social grouping or separation from all other whales—would work for him in captivity.

Tilikum has been valuable to SeaWorld not just because of his one-off appearances as the giant killer whale capable of soaking the spectators in the front rows of the performance stadium. He was central to their national breeding program. He sired more than 50 percent of all of the orcas owned by SeaWorld, despite his record of aggression toward people. The notion that SeaWorld would pick the male most aggressive to humans to dominate its breeding program is as counterintuitive as it comes; you would think that you'd want the least aggressive and most human-friendly orca fathering babies. But breeding orcas in captivity is fraught with problems anyway, and Tilikum handled the task well—originally via copulation and later through artificial insemination. SeaWorld executives elected not to tamper with that record of performance.

Even so, SeaWorld's orca breeding program has been marred by challenges. Some females have rejected their offspring—a circumstance entirely unheard of in the wild. And even when mothers have bonded with their calves, SeaWorld has separated them, causing mothers to cry out in vain to their separated babies for days on end. Cowperthwaite told me that, "Calves being separated from their mothers, and watching how those animals grieve as a result of that, hit people squarely in the gut," as a primary explanation for so many viewers emotionally connecting with *Blackfish*. "The love of a parent and a child is so familiar."

Yet all of the problems with breeding, including cases of incest—also unheard of in the wild—pale in comparison to the capture of wild orcas. In our discussions, Cowperthwaite recounted SeaWorld's history of chasing down pods of orcas in Puget Sound with speedboats and explosives and encircling them with purse seine nets and then plucking out the babies. Orcas died during the chase and capture, while others were left behind despondent that their loved ones had been stolen from them. John Crowe, who captured orcas for SeaWorld, recounted the operation: "The whole family is out here twenty-five yards away maybe . . . and they're communicating back and forth. Well, you understand then what you're doing, you know . . . I just in turn [started] crying. I didn't stop working but I, you know, just couldn't handle it—just like kidnapping a little kid away from his mother." He would later reflect, "this is the worst thing that I've ever done, is hunt that whale." Ultimately, the state of Washington banned these captures, but SeaWorld simply sent its hunt team to international waters, taking Tilikum, Keiko, Kasatka, and others near Iceland.

"We are anesthetized and told to feel happy, with all the music, bright colors, and smiles," Cowperthwaite told me in reflecting on the psychology of SeaWorld and the experience it provides for fans. "You are not doing any critical thinking and you are consuming what they are pouring. You want to have a good time, and it's just easier to go with it."

Cowperthwaite's account reminded me that building a humane economy is not just about businesses doing the right thing, but about consumers doing so, too. It's about driving information to consumers, who act on that knowledge and adjust their behavior in the marketplace. Change comes over time, often after long-running campaigns, with books, investigative stories, lawsuits and lawmaking, scientific papers and commentary, investments by philanthropists, and protests to shine a light on a problem. With SeaWorld, there had been consistent pressure and big cultural events, like the release of *Free Willy* and the release of Keiko, but only fleeting impact. All that emotional and informational capital had been building up, though, and it was not until the death of Dawn Brancheau that the anti-captivity movement would go mainstream. The publication of David Kirby's exposé, *Death at SeaWorld,* primed the issue. But it was Cowperthwaite's documentary that was the tipping point, bringing the mistreatment of marine mammals and the hazards for the trainers into focus. It oriented people who had heard bits and pieces of this story before, but now many

SeaWorld fans pivoted on the issue and started a viral campaign to keep people out of the turnstiles. Long-time critics were emboldened, and SeaWorld was facing an institutional crisis.

"It's about us realizing that they are not that different from us," Cowperthwaite added. "If you come to that conclusion, you then start understanding what it all means to have a mother have her calf taken away from her, and you understand what it would be to need a hundred miles to swim and instead to circle a pool for the rest of your life.

"I had been to the Los Angeles Zoo, and I had a hard time looking at the silverback gorillas," she continued. "I see emotions on their faces, and I see boredom. But with SeaWorld, I had no clue. I didn't see that it was a sad place. Only during the process of making *Blackfish* was I awoken."

Just as the circus will continue without elephants, so amusement parks and aquariums can survive without orcas. They will have to. Of the dozens of aquariums operating throughout the country other than SeaWorld, none keep orcas, and they're more than getting by. Only the Georgia Aquarium seems stuck in its ways, attempting in 2012 to gain permission to import eighteen beluga whales captured from the Sea of Okhotsk near Russia two years prior. (Many of these beluga whales were destined for SeaWorld facilities, according to the import application filed with the government, but in July 2015, SeaWorld reversed its position and said it was no longer in on the deal and would oppose importing the whales.) The US government, under pressure from the HSUS and other organizations, stopped the transfer, maintaining the nation's record of more than a decade without any imports of live-captured whales. In September 2015, in response to a lawsuit filed by the Georgia Aquarium and its partners, a US District Court upheld the National Marine Fisheries Service's denial of an import permit application for the wild-caught beluga whales.

Some aquariums go way beyond avoiding the worst practices. Monterey Bay Aquarium now attracts two million visitors a year, exhibiting animals and plants native primarily to the marine systems of the central coast of California. In addition to providing a glimpse at ocean ecology, Monterey Bay Aquarium rescues animals in need, using its experts and facilities to nurse and rehabilitate sea otters and other wildlife at risk or in trouble. It features a replica of an old-fashioned diner and displays menus that educate people about sustainable and humane seafood choices; videos and other interpretative materials reinforce the lessons. The whole experience is built around the majesty of the ocean in all of its complexity and interdependence, and it's so much subtler and more meaningful than putting a single species on exhibition to perform tricks in a tank. The facility is a central tourist attraction for the city of Monterey, and its brisk attendance shows no signs of letting up.

Monterey Bay may be the leader in the field, but it's not alone in tailoring its work to educate the public about the oceans and about our responsibilities to marine mammals. John Racanelli, chief executive officer of the Baltimore Aquarium, decided to give up the facility's dolphin exhibition when he took over in 2014. He recognized that these big-brained animals simply do not fare well in captivity.

What do we really learn by putting orcas and other cetaceans on display? Mainly we learn that they suffer in being taken from their families, and that they attack each other on a regular basis and their trainers on occasion. That's the equivalent of the circus teaching us that elephants fear the bull hook—a lesson that's obvious and hardly one that's enriching anyone. There's nothing there that stays with us, at least nothing good.

Appointed in 2015 to right the ship, SeaWorld's CEO Joel Manby has the challenge of moving the business away from holding complex marine mammals in small pools and toward giving millions of visitors a more constructive and uplifting message about the oceans. It's so timely, too, especially when we consider the crises facing the world's oceans: the untold tons of trash and plastic polluting the seas; floating nets left behind by fishermen, the overfishing, illegal fishing, and mining of marine resources; and so much collateral damage from energy production and other commercial activities. If he sticks with the old model of keeping orcas for splashy shows, he'll guarantee continuing controversy, protests, lawsuits, a stream of letters from children, and a grinding erosion of public esteem.

In April 2015, just as Manby took the helm, SeaWorld launched a $10 million advertising and social media campaign to convince its customers that its practices are humane. I noticed that CNN stepped up its rebroadcasts of *Blackfish* as SeaWorld stepped up its media buy. Even apart from that countermove, I doubt even $100 million from SeaWorld could turn around the popular perception that it's simply wrong to keep the smartest, biggest animals in the world in such tight quarters. Maybe SeaWorld thinks it can train people to forget what they've learned, by somehow putting a happy face on a story of privation. But in the humane economy, people know corporate propaganda when they see it. In July 2015, after its big advertising campaign, SeaWorld announced its quarterly earnings, and they were down 84 percent compared to the prior year.

Earlier in the year, SeaWorld announced it would increase the size of its swimming pools for the orcas at a cost of $100 million—an immense investment that amounts to a defacto admission that its current orca housing setup is insufficient. It was done clearly to try to demonstrate to the public that the company cares about whales, but one that also signals that the company is making an indefinite commitment to orca shows as part of its business model. In October 2015, the California Coastal Commission, whose members are charged with protecting the state's coastal resources and approving new construction projects, gave the nod to SeaWorld's plan to expand the size of its orca pools, but only on the condition that it breed no more orcas, transfer no orcas from its San Diego facility, and import no new orcas to the state. This action, in an eleven-to-one vote by the appointed commissioners, was yet another rebuke and indicator that social change happens sometimes through the most surprising channels. SeaWorld's stock sank more than 5 percent that day. SeaWorld's expensive proposed facelift boomeranged and turned into an existential threat to its signature location. If the commission action is not reversed, the decision amounts to a sunset provision on its use of orcas in San Diego, given that the shows can go on only as long as the current set of performers can conduct them. California Congressman Adam Schiff followed up by introducing a bill in the Congress to achieve the very same end.

With the pressure mounting, SeaWorld had to act. Its first move was defiant, promising to contest the Coastal Commission action. But then Manby, who had been a highly successful auto executive and had not come up through the ranks at SeaWorld, took a dramatic turn. He pledged to end the theatrical performances with orcas at San Diego within a year—a signal that he recognizes that commanding orcas to splash the customers won't wash away the company's problems. He also told the Georgia Aquarium that he no longer wanted to be a part of taking in belugas that had been captured in the wild. These actions didn't turn around the company's fortunes overnight, but they felt like acts one and two in a series of maneuvers to try to recreate the company and find a business model that works for the future.

I've long been struck by the similarities between Ringling's use of elephants and SeaWorld's exploitation of killer whales. The animals are right at the center of both brands. They are among the biggest, most intelligent, and most charismatic of wild animals. Both companies had run into trouble from regulators based on their animal care and worker safety issues—Ringling with USDA, and SeaWorld with OSHA and then the California Coastal Commission. Both of their industries and their safety records had been tarnished by high-profile incidents that resulted in human fatalities. And both companies had taken babies from the wild to put them on display, and only more recently began to invest in captive breeding to avoid the legal and public relations problems of wild capture.

Yet there the stories diverged, at least initially. The biggest difference now is that Ringling has decided to change its business model, while SeaWorld's instinct has been to build bigger pools. Ringling appears to have turned the corner, but SeaWorld is taking a pounding as it slowly turns the ship. "Ringling said it's not going to use elephants and they invoked our language," Cowperthwaite told me. "The public has decided it's not humane so we are not going to do it anymore. That was amazing for them to show this humility." I've got a feeling that's where SeaWorld must go, too, in the end, since the humane economy is a force that can wear down even the most stubborn-minded companies.

The release of Keiko, in the wake of the success of *Free Willy*, was a wonderful teaching moment for the world, with life mirroring art. It was one thing, however, to win the freedom of a single whale in a small pool at a makeshift amusement park in Mexico, but another to talk about freeing dozens of whales from SeaWorld, a big business with millions still attending its shows. Along with political impediments, there's no doubt that all kinds of logistical problems will present themselves if there's a serious effort to get the whales out of those small tanks. But the naysayers are unlikely to have the final word on this story. This admittedly challenging problem can be solved by using the same determination and resourcefulness that created these holding facilities in the first place. Human ingenuity moved them from sea to SeaWorld, and human ingenuity can set them back where they belong.

"SeaWorld doesn't even begin to speak honestly as a corporation," Cowperthwaite said to me with passion rising in her voice. "Their enterprise is based on taming, riding, and standing on the nose of the animals. They can play up the love part and hug them, and that's genuine from the trainers and people can connect with that. But all they're doing is mastering apex predators."

I asked her what this project has meant for her more broadly, beyond the impact of the film. "It has changed everything," she told me. "I have a deeper reservoir of empathy for all living things. I've been watching these robins build a nest outside our house. I pay attention to the bees and their disappearance in Southern California. My husband and I are doing vegan before 6 p.m. He's an emergency room doctor, and he's loving it."

In 2014, scientists found a 104-year-old killer whale off the western coast of Canada—just north of the very area where SeaWorld used to capture orcas. "Granny," as they called her, is living with her children, grandchildren, and great-grandchildren. When Granny was spotted in 2014, she had just finished an eight-hundred-mile trek from northern California along with her pod—not too bad for an elderly lady. One of her grandchildren, Canuck, reportedly died at the age of four after being captured and held at SeaWorld. The Whale and Dolphin Conservation project estimates that orcas born in captivity only live to four-and-a-half years old, on average; many of the long-living orcas at SeaWorld don't make it out of their twenties.

Granny was born decades before SeaWorld started capturing orcas and keeping them in captivity. She evaded capture and has survived for more than a century as the company has made victims of so many other killer whales. I can only hope that she's still alive when we see SeaWorld end its orca shows and turn its business into a forceful advocate for marine life and the oceans.

"I will remember what I was, I am sick of rope and chain, I will remember my old strength and all my forest-affairs." So wrote Rudyard Kipling in *Toomai of the Elephants*, seeing even in that distant time the dignity of animals taken from their world, and dragged into ours for spectacle and amusement. Now, at least, this appreciative spirit is spreading, and converts are won every day to the cause of helping these animals and treating them with respect. Little by little, we are all growing sick of the sight of rope and chain, along with bull hooks, small tanks, tripwire, and the other archaic tools of the animal entertainment industry. We're remembering what these creatures are, and in every case the reality is more inspiring and wondrous than any show ever staged by man.

CHAPTER FIVE

Animal Testing Yields to Humane Science

"The great advancement of the world, throughout all ages, is to be measured by the increase of humanity and the decrease of cruelty."

—SIR ARTHUR HELPS

A Compact with Our Wild Cousins

In June 2015, holed up in the United Airlines club at Boston's Logan Airport, I called into a press teleconference and heard the trusted voice of my childhood hero Jane Goodall on the line. US Fish and Wildlife Service (FWS) Director Dan Ashe opened the call and announced that his agency would place all chimpanzees on the "endangered" list under federal law. Goodall and I then detailed the many threats chimps face and praised the change in their legal status—the welcome response to a petition filed four years earlier by HSUS, the Jane Goodall Institute, and others. It should never be a cause for celebration when circumstances for an animal are so dire that federal intervention is warranted. But the US government's change in the legal status of chimps marked an inflection point—one species, Homo sapiens, that had done so much harm to its closest kin in the animal kingdom, Pan troglodytes, would now pledge to save that species from further exploitation. For as much pain and loss as we humans inflict on other creatures, we also have the power to assume a new role as their guardians, and who better to defend than our fellow primates?

"This change shows that many people are finally beginning to understand that it is not appropriate to subject our closest relatives to disrespectful, stressful, or harmful procedures, whether as pets, in advertising or other forms of entertainment, or medical research," said Goodall, whose firsthand accounts of chimpanzee communities in Tanzania had been among my first childhood fascinations about animals forty-five years ago. "We are beginning to realize our responsibilities towards these sentient, sapient beings, and that the government is listening." Referring to the animals as "chimpanzee beings," Goodall said the decision "shows an awakening, a new consciousness."

Though not quite the "personhood" designation that some animal lawyers had been seeking, the upgrade in the chimps' protective status remedied an unusual and unlawful "split" listing by the federal government in 1990. That quarter-century-old legal classification had acknowledged the growing threats to dwindling populations by recognizing wild chimpanzees as endangered, but did nothing to protect captive chimps in the United States. This bifurcating of the chimps' status was partly accomplished through a special rule developed just for this circumstance as a political sop to the biomedical research community.

After the 1990 split listing, researchers could still deliberately infect chimps with deadly diseases. Operators of roadside menageries could still put them on display—alone in miserable cages, without even the company of another chimp—as a draw for tourists and passersby. Exotic animal dealers could still breed them and sell their infants across state lines for the pet trade, and occasionally with tragic results. In a case that was both bizarre and chilling, a pet chimp named Travis, sold to a Connecticut woman by a dealer in Missouri, disfigured her friend Charla Nash, biting off her nose, lips, fingers, and toes in an unprovoked attack in 2009. Advertising agencies could even outfit chimps in silly costumes as an attention-getter in television commercials. Such treatment of chimps was part of what motivated Jane Goodall to subordinate her field studies in the 1980s and to start speaking out to audiences around the world. She called attention to disappearing forests and other threats to wild chimps, but also to the plight of the captives in the United States and elsewhere. She argued that conscripting chimps for commercials, while capable of producing a quick laugh, sent the wrong message about how we should interact with these remarkable creatures, undermining global protection efforts.

Dozens of companies featured chimps in ads, and for a time it seemed that adorable apes in tutus would crowd out almost any other form of advertising during the Super Bowl, which has made a separate sport of broadcasting much-hyped thirty- or sixty-second spots. As if the use of chimps were not already an overworked, hackneyed device for advertisers, CareerBuilder in 2012 featured the animals showing off primitive parking skills on their way to the office, clasping briefcases and clad in ties and trench coats. This kind of portrayal of chimps "undermines the scientific, welfare, and conservation goals" of chimpanzee protection programs, wrote Steve Ross, of the Chicago-based Lincoln Park Zoo, who runs the Chimpanzee Species Survival plan for the nation's accredited zoos.

Moreover, acting was never a long-term gig for the chimps. By the age of seven or eight, these wild animals are too powerful and unpredictable to control as actors and can become dangerous. No longer manageable on the set, and too dangerous for picture-taking appearances at a kids' party or at a roadside zoo, the powerful adolescents get dumped into the exotic animal market. If they are fortunate, they wind up in a reputable sanctuary, which, while better than the alternatives, places an enormous long-term, cost-of-care burden on the operators. The chimps are likely to live another fifty years, like Cheetah of *Tarzan* fame, and will be there long after the last current employee retires from the facility.

The 2015 announcement should change the fortunes for most of the 2,000 or so captive chimps now living in the U.S., including 750 in laboratories where life can be unbearably miserable. Their upgraded status won't get chimps out of inhumane forms of confinement overnight, but it is already accelerating their transfer to sanctuaries by creating a new legal standard that private chimp owners must obey. Injuring chimps, and perhaps even separating an infant from her mother, will be legal only if the activity enhances the survival prospects of chimps in the wild. It would be a real stretch for an animal trainer dressing a chimp in some ridiculous costume for an ad, or even a biomedical researcher injecting a drug meant for human use into a chimp, to meet that standard.

The pressure to end the domestic exploitation of chimps had been mounting, for while they are only hundreds compared to the billions caught up in industrial agriculture, the protection of our evolutionary cousins has long carried special moral significance. HSUS and other animal-protection groups threw back the curtain on abuses of captive chimps with undercover investigations at key chimp facilities. Political leaders joined in calling into question the use of chimps and fought efforts by government agencies to transfer more of these primates to facilities for invasive experiments. Advertising agencies, led by the Ad Council, pledged to stop using chimps in ads. Pharmaceutical companies such as Merck and Abbott Laboratories announced they'd halt future use of chimps. All of these efforts got a lift from the emerging consensus that chimpanzees share 98 percent of our DNA and have consciousness, intelligence, and emotional lives to show for it. Leaders of the emerging humane economy developed more effective and cost-efficient alternatives to using chimps in laboratories and exploiting them in television and film. And chimps, too, benefitted from the growing awareness that habitat loss and the bush meat trade are imperiling their existence in the wild. Leading up to the 2015 announcement, it had become increasingly uncomfortable for government scientists and private companies to harm chimps. We are surely headed to a better future when we can look into the mirror without feeling shame for our callous mistreatment of the nearest embodiment of our primitive ancestors.

This has been a long process, and decades ago, I was just one of many young people energized by Jane Goodall's interdisciplinary work in ethology and anthropology. With her first dispatches from Tanzania as a twenty-six-year-old scientist, she altered how we view chimps and added a layer to the story of our own human origins. Working under the guidance of legendary paleontologist Louis Leakey, and disregarding stuffy scientific conventions by assigning names to the chimps and treating them as individuals with distinct personalities, she lived among them in Gombe National Park in Tanzania, after David Graybeard, the best friend of the alpha male Goliath, accepted her. She told the world about the workings of chimpanzee tribal societies—their play and games, maternal care, family squabbles and power struggles, warfare with neighboring clans, and, most famously, their tool use, which had long been considered a defining attribute of humans.

Chimps are not human, but in certain capacities—from problem solving to their truculence with trespassing chimps—these simians reflect some of our own best and worst traits. There was no denying, with their dexterous hands and probing eyes, that they are humanlike. With the up-close accounts from Goodall, the rest of us began to understand the moral implications of these cognitive and physiological similarities. As with many aspects of our dealings with animals, clarity comes only with historical perspective, and typically only after political resistance and endless excuse making from people resistant to change. Only now, half a century after Goodall's work, are hundreds of captive chimps getting a shot at a decent life, and their former persecutors are finally muted, marginalized, or reformed.

A few years before the 2015 announcement, I had traveled down to Keithville, Louisiana, about twenty-five miles south of Shreveport, to visit Chimp Haven. If you're a captive chimp, this is the place to be—a two-hundred-acre facility of open-aired courtyards with natural cover and twenty-foot-tall play structures that allow climbing, swinging, and a 360-degree view of the grounds. There are even faux termite mounds, where chimps use sticks and bamboo to fish out applesauce, ranch dressing, ketchup, and other treats—mimicking natural behaviors. Other chimps there live within actual forests, surrounded by sheer walls and bodies of water (chimps cannot swim, so water features are an elegant enclosure), and they can climb and even nest in trees.

I stood on an upper deck at Chimp Haven and peered down into a courtyard where chimps were playing below, and then straight ahead where three chimps were resting atop structures twenty feet up in the air. I couldn't help but think of the turn of fortune in their lives. Once they were denied meaningful space to roam or even an unobstructed glance at the sun, clouds, or stars; they rarely got out of small, metal cages; and, when they did, it typically meant some sort of pain or trauma awaited them. I threw some fruits and vegetables to the chimps on the platform, and one of them, without a hitch, caught a tomato and threw it right back at me. I snagged it as a reflex and accompanied that action with an open-eyed, open-mouthed look of astonishment, before that expression turned to laughter. He made some sharp movements and some pant-hoot sounds, very pleased to sacrifice a fresh tomato in order to get a rise from me and the rest of the people there.

Animals in laboratories or roadside zoos are known to throw feces at onlookers, but those actions seem born more of frustration or anger than harmless mischief. (Indeed such behavior is not unknown in our own prisons.) They know something has gone badly wrong in their fortunes, and their occasional acts of protest remind us that endless privation is soul shattering for these creatures, no matter the false claims of those who attempt to excuse or explain it.

Established in 1995, after Caddo Parish in Louisiana donated a generous allotment of land, Chimp Haven is a monument to the best of the human spirit, a place created by people of conscience to allow animals who could not be returned to the wild to enjoy the rest of their lives in peace and safety, and with some pleasure along the way. Yet, it's hard to forget that places like this exist only because other people have created a trail of pain, misery, and homelessness for these chimps. The sanctuary is needed only because of the mistreatment and imprisonment that preceded it. Chimp Haven's first president, behavioral primatologist Linda Brent, and other animal advocates developed the sanctuary concept after biomedical researchers bred hundreds of chimps for AIDS work and then realized the animals can contract the disease without displaying its symptoms. The researchers developed no cure and apparently didn't give much thought to all the primates they left behind. It costs roughly $22,000 a year to care for a chimp in a lab, and with each animal expected to live to age fifty or beyond that comes to about a million dollars for the lifetime care of each chimp.

In 1997, the National Academy of Sciences published a report to examine the question of what the National Institutes of Health should do with all these chimps no longer wanted for AIDS research. This report called for a five-year breeding moratorium. It also concluded the government should bear responsibility for the long-term costs of care, dismissed euthanasia "as a general means of population control," and called for the development of a national system of sanctuaries to house retired chimps. Chimp Haven, founded two years prior to the release of the report, was the right place at the right time, and it would become the centerpiece of this network of sanctuaries. The only problem was that the well-meaning animal caretakers at Chimp Haven could not, without an extraordinary effort and luck, raise tens of millions of dollars to care for these chimps until the end of their natural lives.

Brandishing the National Academy's report, the HSUS and other animal-protection groups succeeded in persuading a majority in Congress to enact the Chimpanzee Health Improvement, Maintenance and Protection (CHIMP) Act in 2000. That new law promised to fund a portion of the construction of retirement facilities for government-owned laboratory chimps and to finance some ongoing care. The idea gained traction in Congress because, even then, lawmakers and leaders at the NIH knew that by transferring chimps to sanctuaries they could save money in the long run. A single-species sanctuary could operate more efficiently than a laboratory—perhaps at half the cost per chimp—while offering the chimps much improved lives. Plus, animal welfare groups and philanthropists would pick up a portion of the tab under the financing structure set up in the CHIMP Act.

With the enactment of that law, Chimp Haven soon became officially known as the National Chimpanzee Sanctuary, with its first tenants arriving in 2005. When I first saw the place in 2011, its population stood at 140 chimps, almost all of them retired government chimps previously infected, drugged, restrained, jabbed, and isolated, many of them for decades on end. These chimps were now in a far better circumstance, finally receiving at least some recompense for all they had endured. And the plan was for the sanctuaries to take in hundreds more in the years ahead.

Among the chimps I saw there were a group of five new residents—Jerry, Karen, Ladybird, Penny, and Terry—who had just arrived from the New Iberia Research Center, a biomedical lab about fifty miles away. I felt special pride in seeing this fresh start for these chimps, because just two years prior, the HSUS had conducted an undercover investigation at the facility, which housed more chimps than any other lab in the United States. A selfless colleague of mine got hired at the facility as a technician, in order to give us an authentic inside view of the grim lives of these animals. She worked there for nine months before telling the story, from the chimps' perspective, to *ABC News* and the American public. Ladybird had been at New Iberia for decades, captured as an infant from the wild in Africa (as were her four companions) in 1960; and some of the other chimps our investigator met had been in labs since the late 1950s. Beginning their laboratory-confined lives when Dwight Eisenhower was still president of the United States, here they were in 2011, having known every stage of emotional distress, psychological deprivation, and such a bleak, unpitied existence that some of them had long ago gone mad.

The biggest surprise of the investigation, though, was that most of these chimps were not routinely used in experiments at all. Year after year, they were just being warehoused in cages, with scant attention to their social needs and without any meaningful enrichment. New Iberia was subjecting a small number of chimps to painful experiments of highly questionable value, while keeping a far larger colony of chimps as prisoners on the off chance they might be used for experiments in the future. It was putting the latter group through unending privation for no social benefit at all—just the remote prospect of one. One laboratory director, in an interview for a national media outlet, even compared these chimps to books in a library, saying he could pull one from the shelf as needed. It was no surprise for me to learn, as we publicized the results of the New Iberia investigation, that we were the only nation in the world still experimenting on chimps in laboratories. The Europeans had given it up years before. The African nations of Gabon and Liberia ended the practice. Most countries—Australia, Israel, and New Zealand among them—never went down that path in the first place.

Having grown up inspired by Jane Goodall's work with wild chimps, it was a thrill for me now to see long-held laboratory chimpanzees finally catching a break and ending up at Chimp Haven. After my visit there, I watched a video featuring another group of New Iberia chimps arriving at the sanctuary. With so many of them with gray hair and discolored spots on their faces, the remarkable animals tentatively stepped into one of Chimp Haven's large, open-aired courtyards, feeling grass beneath their feet again for the first time in decades, or, for the first time *ever* in the case of those born in captivity. They felt the warm sun on their backs. No longer were they alone, isolated in the small cages, huddled inside dimly lit buildings. Instead, at last, they looked up and saw the sky, the clouds, the trees, the beautiful world beyond the bleak, manufactured one they thought they'd never escape.

The HSUS's New Iberia investigation, which had prompted the government to release some of the chimps from that facility, added enormously to public understanding of the harsh realities of primate research. The exposure of these abuses and the era of incremental reform began earlier when animal advocates and government officials looked into mistreatment of chimps by the Coulston Foundation, a contract laboratory testing drugs on chimps in Alamogordo, New Mexico. Owner Fred Coulston had earned the reputation as a cold and ruthless character; he was notorious for violating the Animal Welfare Act in his abuse of more than six hundred chimps under his control. In 1999, after inspecting his facility, the Food and Drug Administration (FDA) wrote to Coulston that it had found "serious violations" of Good Laboratory Practice regulations, declaring that "unless these deficiencies are corrected, we would consider future studies conducted at your facility to be seriously flawed." Coulston never turned that perception around, so private companies seeking FDA approval for drug trials stopped hiring him. Then the NIH decided not to renew his Animal Welfare Assurance credentials, eliminating the possibility of Coulston carrying out any federally funded animal research. That one-two punch prompted the Coulston Foundation to declare bankruptcy in 2012, and the eighty-seven-year-old Fred Coulston, a symbol of the old era of callous disregard for animals in laboratories, simply faded from the scene.

Both the saga with the Coulston Foundation and later our investigation of New Iberia made it plain that using chimps for invasive experiments was fraught with a range of very troubling problems. Scientific doubts were growing about the usefulness of chimps in research; more reliable and less expensive alternatives were emerging, and moral concerns could no longer be hidden behind closed doors.

In the case of Coulston, bankruptcy did not end the story. A second major new sanctuary was born to care for the animals caught in the middle—Save the Chimps, in Fort Pierce, Florida. Founded by the late Carole Noon, a visionary animal advocate, Save the Chimps took in more than 250 of the Coulston chimps, after the federal government seized control of the others. But because they were privately held chimps, there was no federal money coming to Noon. Her sanctuary and these chimps needed a savior.

They found one in Jon Stryker, the billionaire heir to a Kalamazoo-based medical supplies company. Stryker told me that as a child he had bought a pet primate, but soon realized that he was in over his head, and came to regret the decision and what this spelled for his primate friend. As a teenager, not seeing many other options, he wrote to Chicago's Brookfield Zoo, asking if the facility would accept a privately owned primate, and to his surprise, the leaders there said yes. He and his family drove the animal to the zoo, and that was the end of his active kinship with a creature who should never have been forced into a life in a human home. The upside of this story, however, is that a child's remorse grew into an adult's activism on behalf of primates. A man of conscience and also of means, Stryker has devoted hundreds of millions of his fortune to helping great apes through the Arcus Foundation—more than any other person has contributed to the cause. He's personally invested nearly $50 million just in the care of the Coulston chimps and others at Save the Chimps—along with several times that amount for conservation of wild great apes. (Stryker and his staff, including Mark O'Donnell, would later play an instrumental role in the advocacy campaign to get all chimps out of labs.)

Carole Noon, Stryker, and the rest of the leadership group at Save the Chimps acquired hundreds of acres for their facility in Fort Pierce. They used earth-moving equipment to dig out twelve islands, where the chimps live with relative freedom, playing in the trees by day and sleeping in houses at night. Each chimp has been carefully introduced to a larger social group for the sort of contact essential to their emotional well-being. There are now more than 260 chimps at this facility, making it the largest chimp sanctuary in the world.

Both sanctuaries saved hundreds of chimps, but another 750 or so remained in laboratories or holding facilities, including 200 chimps that the government seized from Coulston. The NIH had planned to send these chimps to a Texas laboratory for further invasive experiments. But animal groups, led by Animal Protection of New Mexico, the Physicians Committee for Responsible Medicine, and the HSUS, mounted a fierce response. New Mexico's governor Bill Richardson and its Attorney General Gary King, both ardent animal advocates, appealed to the NIH to keep the chimps where they were. Richardson traveled to Washington to meet with top NIH officials to argue his case. "I think it is important to find solutions to hepatitis C, but there are other ways to do it rather than testing on chimpanzees," said Richardson, who was the first elected official to call for a National Academy of Sciences study on the value of chimps in invasive experiments. Richardson pushed for the chimps to stay at Alamagordo, which though not a state-of-the-art sanctuary, ensured they would never be subjected to any invasive experiments.

US Senators Tom Udall and Jeff Bingaman, also of New Mexico, and Tom Harkin of Iowa then added their weight to the cause. Harkin's involvement was especially important because he chaired both the Senate appropriations and authorization committees that control the NIH's budget and activities. The lawmakers jointly wrote to Dr. Francis Collins, the NIH director, and urged him not to transfer the chimps or allow them to be experimented on. They also requested that Collins empanel the National Academy of Sciences to review the status of chimps in biomedical research. "Considering the great progress the scientific community has made in research techniques," they wrote to Collins, "we believe the time has come for an in-depth analysis of the current and future need for chimpanzee use in biomedical research." Collins, who was turning out to be a transformational figure on the broader issue of animal testing as a promoter of alternative methods, postponed transferring the chimps to Texas and empaneled the Institute of Medicine of the National Academies to examine the need for chimps in research.

When the Institute of Medicine came out with its finding nine months later, it was, according to HSUS vice president Kathleen Conlee, a thunderclap. The committee had been charged solely with examining the necessity of using chimps in research—and not the ethics of using chimps. But members declared that this charge was too narrow, finding that "any analysis of necessity must take these ethical issues into account." The committee, whose members selected bioethicist Jeffrey Kahn to run it, concluded "the chimpanzee's genetic proximity to humans and the resulting biological and behavioral characteristics not only make it a uniquely valuable species for certain types of research, but also demand a greater justification for conducting research using this animal model." The committee determined that "most current use of chimpanzees for biomedical research is unnecessary" and recommended curtailing government-funded research on chimps.

Kahn, a professor at Johns Hopkins University's Berman Institute of Bioethics, has said "the bar is very high" for any future uses of these animals in research. The committee acknowledged the use of chimps had aided the development of monoclonal antibody therapies and vaccines to prevent hepatitis C. But it found that alternatives would exist in the near future to eliminate the need for chimps in research—a clarion call to the scientists of the humane economy.

Describing the recommendations as "very compelling and scientifically rigorous," Collins, on behalf of the NIH, announced he had "decided to accept the IOM committee recommendations." He further asked a council comprised of leaders of the NIH's many scientific centers, known as the Council of Councils, to create a working group to implement the findings, including what to do with the thirty-seven or so research projects involving chimps that the NIH was supporting.

Thereafter, the NIH ceased accepting grant applications involving chimps. And in June 2013—eighteen months after the release of the report—Collins announced NIH would retire the vast majority of government-owned chimps to sanctuaries, declaring that "greatly reducing their use in biomedical research is scientifically sound and the right thing to do." His expert panel concluded, "Recent advances in alternate research tools have rendered chimpanzees largely unnecessary as research subjects." The panel recommended phasing out all current biomedical research grants involving chimps in laboratories, ending chimpanzee breeding, and retiring the vast majority of government-owned chimps to sanctuaries. The group determined the federal sanctuary system, run by Chimp Haven, can best meet the needs of the animals. Based on the criteria set forth in the report, not one laboratory could provide an "ethologically appropriate" environment for the chimps. Also under the plan, NIH decided to maintain fifty chimps in the unlikely event they'd be needed in a medical emergency, but they'd have to create a much improved facility to hold them.

The world had already been changing for the better for captive chimps. Previously, the HSUS had campaigned to get pharmaceutical companies to end their use of chimps, and we applied pressure by conducting shareholder advocacy efforts. We secured commitments from Idenix and Gilead. Now the Institute of Medicine report gave reason for others in the sector to rethink the place of chimps in their current and future research programs. "I think it's going to raise the bar for non-NIH-funded research, because federal government standards in research generally, with animal and human subjects, tend to set the tone, to set the pace for everybody," predicted Gregory Kaebnick, a bioethicist at the Hastings Center. And that's just what happened. Within a year, even though the NIH's decision applied only to government-owned chimps, Merck ended its research using chimps.

Abbott, also a pharmaceutical giant, said it would not use chimps in the future and that it "fully supports the findings of the 2011 Institute of Medicine report."

By 2015, Dr. Markus Heilig, a laboratory chief at NIH, would observe that "pretty much everybody has gotten out, or is getting out, of research with chimps." Two of the last three large users of chimps, New Iberia and the Yerkes National Primate Research Center in Atlanta, have pledged to send their 350 or so chimps to sanctuaries. Without government grants for chimp research, and nearly all of the pharmaceutical giants out of the business, there are no clients left. Only the chimps remain—and the expense to care for them.

In November 2015, Collins took another step forward. He and NIH decided to scuttle plans to maintain a colony of fifty chimpanzees for research—reversing the original idea to keep that group in reserve should a medical crisis emerge. Collins chose to lock the door behind the chimps and throw away the key on their way out of the laboratories. Given that no scientist had sought approval for funding to use chimps in experiments in nearly three years, Collins observed, "I think it's fair to say the scientific community has come up with other ways to answer the kinds of questions they used to ask with chimpanzees."

Like so many other successful animal-welfare campaigns, once change gets rolling, it gains a momentum of its own. The actions of government, industry, and animal-welfare advocates become mutually reinforcing. It was the NIH's acceptance of the National Academy's findings that made it much easier for the US Fish and Wildlife Service to later list captive chimps as "endangered." The wildlife agency faced rising pressure from an increasingly knowledgeable public but hadn't wanted to tangle with pharmaceutical companies and the biomedical industry on Capitol Hill on the issue until the companies and the NIH realized that continued research on chimps was neither right nor necessary.

The dynamics of the humane economy will assure that better and more cost-effective alternatives to animal research prosper. That leaves the task of moving remaining chimps from laboratories to sanctuaries and caring for them—an immensely costly proposition, but one that the government, private philanthropists, and charities must undertake.

One day soon, we won't see any chimps in labs, or in commercials, or in private owners' backyards or basements. Most advertising agencies—from Young & Rubicam, to Ogilvy and the Grey Group—have committed to no longer using great apes in their ads, so we're unlikely to see them appear during the Super Bowl, except perhaps in computer-generated form. That ad agency outreach, conducted principally by People for the Ethical Treatment of Animals, is yet another indicator of the humane economy at work. As with so many reform efforts, the pressure comes from many angles. And if we play our hand right, we'll see more of them in the wild, living free, among their families, and using sticks to probe real termite mounds. As in most things, nature does the upkeep, and it's authentic and good. So far as captive great apes are concerned our conscience will be clear, and in a humane economy that counts for a lot.

Animal Testing Disrupted

Not long after the explosion and blowout at the Deepwater Horizon rig killed eleven workers in April 2010 and roiled the waters of the Gulf of Mexico, I joined US Senator David Vitter in his home state of Louisiana to help animals at risk. In an oil spill, unlike in a hurricane where wild animals often display a sixth sense and flee to safety ahead of the path of the storm, marine creatures don't seem to understand what's happening until it's too late. Plumes of oil, pushed by wind and current, are a silent and creeping menace that blotch fur, feathers, skin, and scales, and at times even coat a victim from head-to-toe or head-to-tail. With all their survival skills, these poor animals have no good means of shedding the glue-like crude—a slow killer.

"This is Barataria Bay," said Vitter, his voice crackling through my headset even though he was buckled in right next to me as the vibrating blades of the helicopter lifted and pulled us over vast stretches of marshland reaching out through Jefferson and Plaquemines parishes and extending into the Gulf. Vitter and I wanted a firsthand look at the search-and-rescue, bird cleaning, and oil containment strategies, and we wanted to make sure animals were a priority.

For decades, Louisiana has been yielding marshland to the ocean at an extraordinary rate—the equivalent of a football field of it every hour—and the Deepwater Horizon spill would compound the problem, infiltrating and degrading these sensitive areas. If oil then burrowed deep into the marshes and remained—as studies had shown it had in Alaska's Prince William Sound after the *Exxon Valdez* oil spill—it could poison wildlife and cause mutations in their offspring.

"I cannot believe how much oil is in the water," I barked into my headset.

Vitter and I wondered whether the efforts from the army of spill responders and nature's own army of microbes could contain the damage and allow a functioning marine ecosystem to return soon. We were mildly relieved when we didn't find much visible oil at Grand Isle, the only human-inhabited barrier island in the state. We landed and then boated to Queen Bess Island, a rookery teeming with life, but with ominous signs of the spill. A bright orange blocking boom and an inner concentric white absorbent boom ringed the island. Vitter observed that the orange boom looked as if it had been covered in "chocolate syrup" and the white boom bore no resemblance to its original pale coloring. Far worse, despite the placement of the booms, the rocky shores of the island were already stained. While we were there, a government boat pulled up and workers reeled in the oil-soaked booms and encircled the island again with clean ones, in what seemed to be a losing battle against a rising and spreading tide of oil.

Later in the day, we returned to Plaquemines parish to see a triage center for oiled birds. As we arrived, wildlife rehabilitators had just taken in about half a dozen brown pelicans, each mottled in dark black crude. Treated birds had been recovering in tents, after workers in bibs and gloves gently but persistently scrubbed them with Dawn dishwashing liquid. There were about 450 or so birds in the animal MASH unit, and officials told us that fifty or sixty more arrived each day. The rehabilitators would then shuttle the cleansed birds to the eastern reaches of Florida for release, in the hope they wouldn't fly back to the polluted waters of the Gulf.

The next day, we headed to the Audubon Nature Institute, affiliated with New Orleans's zoo. The facility had taken in fifty oiled turtles—most of them endangered Kemp's ridleys. The center had coordinated with first responders to call in coordinates of turtles in trouble so that trained personnel could retrieve them. It's one thing to pick up an oiled bird and another to retrieve a turtle that might weigh one hundred pounds. Yet even with these systems in place, personnel at the center there feared what they could not see—untold numbers of the prehistoric-looking creatures languishing in the marshlands without anyone's noticing their despair.

Even though I knew we'd find vast numbers of creatures in distress, the up-close images remain hard to clear from my head years later. Something I didn't expect, though, was that my time in the Gulf gave me a window into an another animal issue entirely: the debate over animal testing, and the role it plays in risk assessments for chemicals found in a vast array of products in everyday commerce, including those used in oil spill responses.

This convergence of the old economy (in this case, the oil industry and the animal testing sector) and its hazards (spills, pollution) with the emerging humane economy and its breathtaking promise was born of urgent happenstance. In the Gulf Coast deployment, the federal government treated the oil with chemical dispersants as a way to break down the crude and allow ocean-dwelling bacteria to detoxify what remained. These commercial chemicals themselves had undergone very limited risk assessments as an environmental hazard, so nobody quite knew the consequences of releasing them in enormous volumes and with indefinite staying power in a marine ecosystem filled with millions of creatures of so many types. While Vitter and I both wanted to mitigate the effects of the oil on wildlife, we worried that the cure could create a new kind of curse.

Through the years, Senator Vitter and I had partnered to pass federal laws to make it a crime for anyone to attend a dogfight or cockfight, and to compel the US Department of Agriculture to license and inspect Internet sellers of puppy mill dogs. At the time of our Gulf trip, we hardly realized our experience there would give us a practical grounding for this different task: creating a new regulatory standard to upend our nation's reflexive reliance on animal testing for examining the safety of chemicals, and, instead, establishing alternative methods as the new industry standard.

The HSUS has pushed for alternatives to animal testing since our founding in 1954, principally because of the moral cost of using animals by the millions in tests. The idea of laboratory animals standing in for humans was so universal that it found its way into common usage, with people casually saying they didn't want to become "human guinea pigs" or "lab rats." Animals have long been utilized as supposed "canaries in the coal mine" to test for toxins. In medical schools, and in medical practice, we widely practiced on them before we operated on people. In car crash testing, and even space flight, we put them in the driver's seat. Indeed, before the first human went into space in the 1960s, we sent dogs and chimps in unmanned spaceships as our surrogates.

For the right reasons, such protocols were established with the goal of providing answers about the human safety of drugs and chemicals. Eventually, animals became the primary and ever expanding tools of choice for industry and government safety testing, while not nearly enough thought was given to potentially better and more humane strategies. After a drug used to treat streptococcal infections and labeled "Elixir Sulfanilamide" poisoned and killed more than one hundred patients in 1937, Congress passed the Food, Drug, and Cosmetic Act of 1938 to require "evidence of drug safety before marketing." In 1962, Congress also expanded drug testing for safety and efficacy, in the wake of the thalidomide disaster—a morning sickness drug that caused women to birth children with deformed or missing limbs.

In the years since, Congress has added safety testing requirements for food additives and pesticides. In 1976, just years after the environmental awakening in the late 1960s and early 1970s and after Rachel Carson's *Silent Spring* and the toxic Cuyahoga River in Ohio literally caught on fire, Congress enacted the Toxic Substances Control Act. This Act empowered the newly created Environmental Protection Agency to assess the risks of chemicals to protect people—and that meant thousands of products used in the marketplace. For tests of food additives, drugs, cosmetics, pesticides, and chemicals, scientists poisoned animals with high dosages of the substance. While it would have been impossible not to glean valuable insights into the human condition by using so many animals so routinely, it also became clear that animals were not mini-humans. And, in fact, test results were sometimes inconsistent even between mice and rats. Thalidomide underwent animal testing prior to its first test marketing on pregnant women suffering from morning sickness, yet these trials didn't provide warning of its horrific side effects—severely enfeebled and deformed babies. After the hepatitis drug Fialurdine had been tested on mice, rats, dogs, monkeys and woodchucks, it was used in a clinical trial in 1992 and led to the death of five of fifteen human volunteers who signed up for it. In 2006, a potential new drug for a type of leukemia as well as rheumatoid arthritis was given to six volunteers, and "after [the] very first infusion of a dose 500 times smaller than that found safe in animal studies, all six human volunteers faced life-threatening conditions involving multi-organ failure for which they were moved to [the] intensive care unit."

And there are plenty of drugs and foods that are beneficial or benign to humans, yet poisonous to animals. Imagine how diminished our lives would be if we had first tested chocolate on dogs long ago and determined that it was toxic. Or consider aspirin, which is poisonous to many animals. If today's regimen of animal testing had been employed back then, aspirin would never have been approved for humans, let alone be sold routinely over the counter. The sweetener saccharine causes bladder cancer in male rats but does not affect humans in the same way.

In 2004, FDA reported that 92 percent of new drugs failed human safety standards, despite passing batteries of animal tests. Since that 2004 report, this pattern has continued. When something doesn't work nine out of ten times, it hardly seems like a very valuable tool, and certainly not one to be used in a life-and-death situation. Animal testing has provided a psychological salve for us in toxicity testing, allowing us to believe we are safer; but it simply has not provided us with a practical means to assure human safety. Instead, it has become a useful tool at times for corporate lawyers. Rather than provide real safety protection for consumers, the reports of animal tests serve as legal defense for manufacturers and distributors. In a personal injury class action, animal testing allows the manufacturer to argue in court that the questioned substance was studied and the results found no conclusive problems of harm.

Today, it's that extraordinary failure rate along with the enormous expense of using animals in time-consuming protocols that are causing forward-thinking scientists and pharmaceutical companies to find better ways to test drugs, chemicals, and other substances. The cost of government regulation isn't driving this by itself either. Scientific entrepreneurs were simultaneously racing toward that sweet spot where profits awaited those who could cut animals out of a top-heavy and ineffective testing process. It's a system that's been long overdue for economic disruption.

When the government needed answers about the effect of the dispersants released into the Gulf of Mexico in 2010, it had no choice but to look to alternatives to animal testing methods, since the exigent circumstances called for a tool that could produce results in a much shorter period of time. The Deepwater Horizon blast had dislodged drilling equipment, resulting in black crude erupting from the sea floor a mile below the ocean surface at an astonishing flow rate.

British Petroleum first told the press the damaged offshore drilling unit was spewing out one thousand barrels a day, then five thousand; by the time I arrived, the company conceded it could be seventy thousand. Senator Vitter ominously described it as "a new oil spill every day." Even accounting for the fifteen thousand barrels recovered each day, that's more than 4.7 million barrels over the blowout's eighty-seven-day duration. That's two hundred million gallons that went uncollected—twenty times the volume that tarred 1,300 miles of coastline and 11,000 square miles of ocean in the *Exxon Valdez* spill. It affected a surface area about the size of the state of Oklahoma.

The spill had a vertical impact as well, since the oil poured out at the sea floor and billowed to the surface, affecting marine creatures up and down the water column. Of course, the most visible impact came to creatures who, like us, live at the intersection of the ocean and the atmosphere and the sea and the coast. Dolphins washed ashore dead, and birds with badly darkened plumage vainly preened themselves—an instinctive behavior that resulted in oil coating their insides, too.

Stopping the gushing of oil was the first priority for the government and for BP, but right up there was containing it and then gathering it up or dispersing it. When Vitter and I got in the air, we looked down upon an armada of ships that was vacuuming oil from the surface. But it had to be slow-going work; it wasn't quite bailing the ocean with a teacup, but the volume of oil and water easily overwhelmed these floating skimming-and-filtering operations. Every gallon removed did some good, but the government decided that what remained in the water would need to be dispersed. And this would be done with chemicals.

Under the 1976 Toxic Substances Control Act (TSCA), the EPA has long been charged with preventing an "unreasonable risk of injury to health or the environment" from tens of thousands of chemicals in commercial use. Lead has been a known toxin for two thousand years, but in the 1960s we became more alert to the effects of DDT, PCBs, and other chemicals that proved to be a very toxic alphabet soup. Unfortunately, TSCA, right from the start, proved grossly deficient in protecting the public from the dangerous chemicals. The original law grandfathered the chemicals in use in the marketplace at the time of enactment. And in the intervening forty years, the EPA conducted comprehensive assessments of only two hundred chemicals and succeeded in restricting the commercial use of only five classes of chemicals—PCBs, chlorofluorocarbons, dioxin, asbestos, and hexavalent chromium.

The safety screening regimen, heavily involving the use of animals, isn't even close to adequate. In the four decades since the enactment of a statute to address the problem, we've been flying blind when it comes to a storehouse of potentially hazardous chemicals that touch our lives. Today, the EPA has an inventory of 85,000 chemicals, with 2,700 used at high volumes, and modern chemistry is introducing 1,500 chemicals a year for possible use in the marketplace—for cleaning supplies, food packaging, furniture, cosmetics, plastics, toys, and everything you can think of. In so many cases, we hardly know these chemicals surround us until we focus on the problem or until we learn of a cancer cluster or people getting some other illness. We've needed less focus on the animal tests and more focus on allowing fewer chemicals in our midst.

In the Gulf, something needed to be done, and far more quickly than the creaky old animal testing system could accomplish. With everyone seeing images of the disaster response on television, including the use of nearly two million gallons of dispersants dropped by low-flying aircraft, the questions could not be ignored—not from regular Americans, the press, or from high-ranking public officials, including Senator Vitter. Would these chemicals cause cancer, act as endocrine disruptors, or cause birth defects in people? Would they do the same to fish, and would the substances accumulate in their tissue and then land on our dinner plates if these fish made it into the food supply? How long would the chemicals persist in the environment, and would they affect future generations of people too? Plenty of scientists have long questioned any large-scale use of dispersants because they're thought to retard the work of oil-eating ocean bacteria and just push oil down beneath the ocean surface. But given that dispersants were part of the plan for the Gulf spill response, it was critical to understand the safety of the dispersants and other chemicals about to be used in enormous volumes.

Scientists would have a hard time offering definitive assurances if they had all the time and money in the world, given the blend of chemicals and species, and an indefinite timeline where problems could erupt. In this case, we had an active crisis and a very narrow window for a risk assessment, which would inform the on-the-ground response. Animal tests, which require administering high doses of a substance to large numbers of mice, rats, or rabbits, sometimes over multiple generations, and measuring organ failure, carcinogenicity, and other harms, were in this case easily disqualified because they were unworkable in the required time frame. Animal toxicity studies can take three to five years per chemical, costing several million dollars per chemical, to say nothing of the costs to the creatures conscripted. A single evaluation of a potential carcinogen requires more than four hundred rats and four hundred mice for a five-year experiment, carrying a price tag of $4 million.

Instead of conducting pointless animal testing, the EPA's Office of Research and Development turned to faster methods of toxicity testing—employing high throughput assays that rely on robots and small concentrations of the chemical placed in testing racks that measure the effects on human cells at different concentrations. The EPA did the lab work and then published a report on eight dispersants in weeks, rather than months or years, providing some practical information to emergency response managers long before the last dispersants had been dumped in the ocean. In the end, the EPA found that some of the products could be endocrine disruptors in fish, but found the most commonly used dispersant to be the least troublesome.

In the years since the Deepwater Horizon spill, US Senators Tom Udall, D-N.M., and Vitter focused on revamping the TSCA and creating a far more rigorous regulatory scheme for safety testing of chemicals. They carried on the original work of Senator Frank Lautenberg, D-N.J., who passed away not long after he made a crusade of the issue. It was clear the nation could do better over a span of four decades than to test a few hundred chemicals out of tens of thousands on the EPA's backlogged inventory. Plus, there was ambiguity in the law, a long-standing free pass to the chemical industry for tens of thousands of its potentially hazardous substances already in the market. The answer was not to continue a prohibitively costly, crude, and ineffective primary safety assessment tool that relied on poisoning generations of animals with concentrated doses of single chemicals that humans would almost never encounter in a real-life situation.

An important pivot point in the toxicity testing debate came three years before the Deepwater Horizon disaster, when an expert panel of the National Research Council convened at the EPA's request concluded that the future of toxicology would rely mainly on testing methods without animals. "The new approach would generate more relevant data to evaluate risks people face, expand the number of chemicals that could be scrutinized, and reduce the time, money, and animals involved in testing," announced the NRC in releasing the report. The new methods involved fanciful-sounding things like bioengineered organs-on-a-chip, robot-automated high-throughput human cell and gene tests, next-generation computer modeling, and modular in vitro human immune systems. Far from fanciful, these methods and others are currently transforming safety testing around the world by enabling scientists to study how chemicals react in the human body at the cellular and molecular level. They are, according to top scientists and regulators, vastly improving the quality of research and offering the potential to replace animal testing altogether. Plus, they allow scientists to test thousands of chemicals at various concentrations in just a few weeks.

Following the publication of this report, members of the expert committee spoke about their recommendations at many subsequent toxicology meetings. Initially, they were met with skepticism and outright rejection from their colleagues in the field. Before long, though, others began to grasp the potential of the new technologies. Beginning in 2013, the Society of Toxicology organized three "Future Tox" meetings to explore the committee's report and how its recommendations might be implemented. The prevailing concern among most toxicologists now was not, "Can we replace animal testing?" but "When it will occur?"

Other nations had also assessed the risk of chemicals and recognized that animal-testing models simply couldn't be useful in dealing with the backlog of chemicals in use. While the EPA started the focus on rapid non-animal testing, Canada was the innovator in using these tools for chemical safety assessment on a large scale. Canada required its regulators to examine a backlog of twenty-three thousand existing industrial chemicals with deadlines too strict to allow reliance on animal testing. As with the situation at Deepwater Horizon, this forced regulatory authorities to make an immediate switch to available computer models and other non-animal tools to sort the chemical backlog according to relative levels of concern.

Europe was also grappling with the challenge of evaluating the human safety of a huge number of chemicals, looking at thirty thousand substances through its new chemicals law. While the EU's tackling of this problem was imperfect, it created a new standard to require the acceptance of validated alternatives to animal testing. It also set the stage for the adoption of future methods and reducing testing by encouraging companies to work together to submit information. As they were developed over the years, HSI collaborated with US and EU regulatory authorities and industry to win acceptance of alternative methods to lethal dose animal tests, sparing fifteen thousand or more rabbits and rats; replaced a wasteful animal test for reproductive toxicity, preventing the poisoning of up to 2.4 million rats; virtually eliminated rabbit eye and skin irritation testing, saving about 21,000 rabbits; and paved the way for full replacement of mouse and guinea pig tests for skin allergy, sparing up to 218,000 animals. Animals and humans were both winners—the humane economy finally being realized after regulatory and scientific foot-dragging for far too long.

From 2010 to 2012, HSUS and its international affiliate worked with European institutions and companies to revise testing requirements for pesticides, which had required the use of ten thousand or more rodents, rabbits, dogs, and other animals for every new pesticide chemical registered for sale. Imagine row upon row of dogs in cages, forced to consume toxin-laced food every day for a year, growing sicker over time, until they are killed for dissection; or rabbits locked in neck restraints while a pesticide chemical is dripped into their eyes or onto the shaved skin on their backs. It was such an archaic and horrifying method, but we won acceptance of an alternative method that involved new technologies, rather than the long-standing poisoning of animals.

Now it was up to the United States to make some progress on the chemical legislation front. With the leadership of Senators Udall and Vitter, and a determined and deft effort from Senator Cory Booker and his staff to demand the use of alternatives to animal testing, the Senate approved a new standard for safety testing of chemicals by a unanimous vote. Animal testing would no longer be the primary tool, but the last resort, under the terms of their measure for both existing and new chemicals. When it's ushered into law, it will consolidate information, require actual safety testing, and allow the use of animals only in rare circumstances. It wasn't as if the ethical concern about animal testing was the singular driver of reform. It was also the lack of reliability and the cost and time to test a single chemical. The old system of testing chemicals was badly broken, and we as a nation were overdue for a reset.

Throughout the post–World War II era and during the emergence of toxicity testing on animals as the primary means to measure risk, HSUS had worked—and over time rather successfully—to win the scientific establishment's acceptance, at least in principle, for the three "R's" of animal testing: *r*efining techniques to minimize pain and distress to the animals, *r*educing the number of the numbers of animals used in testing protocols, and *r*eplacing them with non-animal methods where available. In 2000, animal-protection organizations, industry, and other stakeholders won approval of legislation to create a government body, known as ICCVAM, to "establish, wherever feasible, guidelines, recommendations, and regulations that promote the regulatory acceptance of new or revised scientifically valid safety testing methods that protect human and animal health and the environment while reducing, refining, and replacing animal tests and ensuring human safety and product effectiveness."

This law itself validated the view that there are moral problems associated with using animals and that they should not be used gratuitously—and not at all if alternative methods are available. Fifteen years later, with a series of extraordinary innovations in technology, we've observed a meaningful reduction in certain categories of animal testing, with the growing (if somewhat grudging) recognition that non-animal methods have become faster and more cost effective than the conventional animal-based toxicity tests. Here again, both ingenuity and practicality are helping to grow the humane economy, but as in all inflection points, the transition from the old order to the new requires consumers, policy makers, regulators, scientists, and corporate leaders to step decisively into this new realm and leave old ways behind.

Those who had been part of the old paradigm for testing of chemicals and drugs were now rethinking it—on both risk assessment (to avoid making someone sick) and efficacy (the ability of a drug to help a patient). Dr. Francis Collins, the NIH director who was so instrumental in ending the use of chimps in invasive experiments, has spoken out about how animal tests for drug safety aren't doing us much good. "With earlier and more rigorous target validation in human tissues, it may be justifiable to skip the animal model assessment of efficacy altogether," announced Dr. Collins. His predecessor at NIH, Elias A. Zerhouni, had similar things to say. "We have moved away from studying human disease in humans," he said, in discussing the reliance on animal testing. "Researchers have over-relied on animal data. The problem is that it hasn't worked, and it's time we stopped dancing around the problem . . . We need to refocus and adapt new methodologies for use in humans to understand disease biology in humans."

When we hear about television commercials for a variety of prescription drugs—whether for depression, erectile dysfunction, rheumatoid arthritis, pain related to nerve damage, and others—my wife, Lisa, and I are amazed at the range of side effects articulated. You typically see a well-groomed middle-age couple smiling, with an idyllic setting, as the narrator proceeds to tell you about "suicidal thoughts," "hostility, agitation," "swallowing your tongue," "skin reactions, some of which can be fatal," and so many other horrible outcomes. If the animal tests were used in these cases, and were reliable or predictive, were they really protecting us or just signaling that we are all rolling the dice when we take these products? And if these warnings do produce these kinds of side effects, are we really just swapping one set of problems for others?

I decided to go talk with some of the scientists at the cutting edge of this revolution in the laboratory, and who have no specific commitment to reduce or end animal testing but want only to make testing more reliable on safety and efficacy grounds. I went to see Dr. Chris Austin of NIH's National Center for Translational Sciences, which began its work in earnest in 2004 to "turn observations in the laboratory, clinic, and community into interventions that improve the health of individuals and the public." The work is broad ranging—from looking at rare and genetic diseases to accelerating the approval process for drugs—but when it comes to safety assessments, it is looking at "testing 10,000 drugs and environmental chemicals for their potential to affect molecules and cells in ways that can cause health problems." With his robots and in vitro tests, Dr. Austin can test tens of thousands and even millions of chemicals—an entirely different order of magnitude from animal testing. "Every week we test 500,000 chemicals at multiple concentrations," Austin said. "We do three million tests of those chemicals every week in a different system, so frankly the collection of backlogged chemicals is like pencil dust." Supported with hundreds of millions of dollars from the federal government, which is by far the world's biggest financial backer of research and development, Dr. Austin is looking at the problem from an entirely different vantage point, and using twenty-first century tools that give regulators more practical insights into how chemicals and drugs affect human health. Again, that high-impact work demonstrates that scientists and government have a big role in the humane economy, especially when it comes to basic research and testing.

And it's not just an issue of dealing with the backlog and demonstrating an ability to test tens of thousands of chemicals in a timely and cost-effective manner—it's also that the animal tests just aren't working. "One can lay a lot of the clinical failures at the paws of the animal data," said Dr. Austin. He noted that half of preclinical animal toxicology results aren't borne out in people. And 80 percent of Phase II trials, those involving larger numbers of people, show no value.

He's quick to point out that he and his staff are only looking at the cellular level. What's lost in this process is what's happening to organs or the whole body. But he's got an answer for that. "The complementary approach is the tissue chip approach." These technologies allow researchers to create simple organ and tissue mimics from human cells (and those of other species) that function much like they would in a living organism ("in vivo"). These "organs-on-a-chip" can be used to study disease or the effects of chemical or drug exposure. Austin and his team are working on these chips, but they're also being developed and improved in other laboratories. The US Defense Advanced Research Projects Agency (DARPA), along with the NIH and the FDA, recently invested $70 million to develop up to ten organs on chips, and European governments and companies have invested billions in the new Innovative Medicines Initiative. These approaches are leading to a world of personalized medicine that may eventually allow substances to be assessed for their impact in my liver as compared to my wife's.

Dr. Don Ingber, who runs the Wyss Institute at Harvard University, hosted me and a colleague to discuss his organ-on-a-chip work, where he is developing livers, hearts, and other organs outside of a living animal and without needing to hurt or sacrifice any living being. His lung-on-a-chip mimics the mechanical functions of a living, breathing lung, and he enthusiastically clicked on a video to show how he and his team constructed this device. He is creating new structural designs that can replicate many of the effects of a living system, to get a better read on how perturbations occur in organs and other biological pathways. "We were getting levels of functionality that no one has ever seen before," he told me in a conference room at Harvard. "These tools are as good or better than an animal model in terms of mimicking a disease process or a function." Ingber told me that pharmaceutical companies, the FDA, and other agencies that have long relied on toxicity testing in animals are excited about the new methods and using them to do better science and to make better and sounder safety assessments. It's no accident Harvard has shut down its primate research center, acknowledging in the process that primate research isn't part of the university's future plans.

There's also a revolution underway in the testing of cosmetics—also an immense industry, like the chemicals sector, that touches every one of our lives, but often in imperceptible ways. For decades, cosmetics manufacturers poisoned animals by the millions, because that was the acknowledged standard and would be accepted by the courts as a legal shield in case they were sued. In the United States, there was no specific requirement by the federal government that cosmetics companies use animals in testing, but it was the standard approach for drugs and chemicals, so companies used it for cosmetics too. In one of the primary tests, the "Draize eye irritation test," the technician placed a high concentration of a compound in the eye of a rabbit and attempted to assess the redness and tissue destruction on a 110-point scale. But the structure of the cornea of the eye of a rabbit differs significantly from that of a human. Rabbits also produce a smaller volume of tears than humans, allowing chemicals and other irritants placed in rabbit eyes to linger longer and cause more irritation. Not only does this make the Draize eye test likely to overestimate the human response, but it also adds to the immense suffering of the rabbit. Unsurprisingly, laboratory technicians typically hated performing the Draize test.

But now, more than six hundred cosmetics companies have done away with animal testing by using new technologies and relying on compounds and substances already judged to be safe. John Paul Mitchell Systems, led by John Paul DeJoria, was among the first, and was joined by Anita Roddick at the Body Shop. Once pioneers, those corporations now face plenty of competition from LUSH and a cascade of other companies that are also making the pledge to deliver products known to be safe for consumers and that don't require rigorous safety testing. The beauty- and fragrance-products conglomerate COTY, which bought the beauty division of Procter & Gamble in 2015, has endorsed the federal Humane Cosmetics Act, a new legislative initiative to end using animals in safety testing for cosmetics. These companies proudly wear the "no animal testing" phrasing and symbols on their packaging. Can you even imagine that any company still testing on animals would proudly state that "we conduct animal tests?" The companies conducting such tests are becoming outliers, for the time being relying on the public not knowing about this kind of cruelty in their research programs.

But they're unlikely to get away with that conduct for much longer, as lawmakers and regulators in multiple nations are revamping the legal framework to forbid these tests under any circumstances. A European Union regulation went into effect that bans all cosmetic testing on animals and also forbids the sale in Europe of cosmetics tested anywhere in the world. This policy kick-started industry and government regulators to drive for the adoption of alternative methods. A year after the EU adopted its ban of cosmetics tested on animals, India adopted a similar law. Together, these laws have closed markets for 1.7 billion consumers to multinational cosmetic and toiletry manufacturers that insist on testing on animals. For years, China resisted the movement by requiring animal testing for all cosmetics sold in the country. But in 2015, China lifted that requirement for cosmetics manufactured domestically. (Animal testing must still be performed for cosmetics manufactured outside China.) A number of cruelty-free cosmetics companies had lobbied the government and urged them to end this archaic requirement, including John Paul Mitchell Systems, which said it wouldn't sell cosmetics in China if it had to conduct animal tests.

Today, with so many companies selling throughout the world, it's unlikely you'll see two tracks of research and safety testing, and over time more companies will conform to the non-animal testing methods. L'Oréal made headlines in spring 2015 when the cosmetic giant announced plans to begin producing substitutes for human skin with 3-D printers. Scientists have been able to farm artificial tissue of human skin cells in laboratories for years. These artificial skin models are used, among other things, to test the safety of ingredients in foundations and eye shadows. And they have proven more accurate at predicting skin reactions in humans than tests performed on animals.

The new non-animal methods offer the prospect of actually testing thousands more chemicals, simplifying cumbersome approval processes for cosmetics, and increasing confidence for corporations and consumers in selling and buying products in the marketplace. If these methods are anywhere near as successful as the Human Genome Project, we can expect big economic returns. The Battelle Institute produced a report in 2011 that looked at the Project's impact and reported that the $5.6 billion spent on the Project (in 2010 dollars) led to $796 billion of new economic output from 1988 through 2010. That's a return on investment of 141 times.

The prospects for a similar return on investment for a Human Toxicology Project are probably just as promising. Innovate UK recently produced a non-animal technologies roadmap for the United Kingdom that highlighted the promise of new non-animal technologies for the pharmaceutical sector. The global pharmaceutical industry generates net income of $980 billion a year but it takes $1.8 billion to produce a single new drug—with the majority of drugs failing during development and clinical trials. The report notes that the market potential for cell-based assays in drug discovery and safety assessment could reach $21.6 billion by 2018.

And there's a very human dimension to all of this. Whereas the EU has been subjecting about fifty chemicals to safety testing every year with animal testing, the new technologies have the potential to subject all thirty thousand of the target chemicals to a comprehensive set of high-throughput human cell and other tests in just one year. And the alternative tests could do so at fifteen different concentrations. (Animal tests typically use just two concentrations of the test chemical.) The data generated in these tests could then fuel a rapidly expanding analysis of predictive algorithms that would identify patterns of behavior among these chemicals and allow us to make better predictions (than the animal tests) of human safety and risk from those chemicals.

In the end, when we move decisively toward this new approach to chemicals testing, we'll be making progress as a society. We'll have greater and more reliable economic activity in the research and development stage and more confidence at the retail stage. Consumers will feel better and more assured about the safety of products they put on their skin or into their bodies, and we'll be able to better react when there is a disaster such as a chemical spill or oil blowout that impacts our drinking water or environment. And we will not have induced so much trauma and suffering in countless creatures who never deserved this dose of inhumanity.

CHAPTER SIX

The Visible Hand and the Free Market: Humane Wildlife Management in the United States

Wolves and Alpha Returns

Coming back to Isle Royale was a reunion of sorts—not the kind where you reconnect with people and a place, but in this case just the place itself. I had returned to this wilderness where I had left a few fleeting footprints decades ago, but it was the wilderness that had left the permanent mark on me.

I had first laced up my hiking boots during a summer thirty years ago on the island—among the most isolated of America's national parks and buffeted on all sides by the cold swells of Lake Superior. I'd been drawn to the park's wolves, moose, and boreal forest since reading about them in *National Geographic* magazines and books as a kid, and I had long dreamed of seeing the place firsthand. So when I got my acceptance letter from the Student Conservation Association for a summer position there, it was a rush not unlike getting into my college of choice. Sound the trumpets! I was on my way to Isle Royale, with backpacks, wolf packs, and much more awaiting me. I could hardly wait to step on ground freshly trod upon by wolves and hear their songs in the stillness of the night.

That was May 1985. There was a sharp chill in the air, and the waters still held onto the subarctic cold from the winter when I arrived by ferry from Houghton, on the north shore of Michigan's Upper Peninsula. After two weeks of training that took me to all major parts of the islands, I was assigned to patrol the island on foot and by boat, to interpret the wonders of the park for visitors, to pull duty at the visitor center, and to perform assorted other tasks that gave me plenty of angles for a proper introduction to this epic landscape. The main island and the outlying ones that make up the archipelago had no outsized features—no towering mountains, ancient tree species, or cascading waterfalls. Yet their rocky shores, crystalline lakes, and aspens, spruces, and firs left any alert visitor in wonder at the beauty of it all.

Isle Royale's wolf pack started with emigrants from Ontario in the late 1940s, trekking in winter some twenty miles across ice-covered Lake Superior. It wasn't a paradise for these lupine pioneers, who were used to punishing winters but perhaps not to the unyielding snows. But there were plenty of reasons to stay. There were no permanent human settlements on the island, and the people who visited came and went without any guns on their hips, leghold traps slung over their shoulders, or livestock in tow. Congress had restricted hunting, trapping, and grazing on Isle Royale in 1940 when it designated the 210-square-mile archipelago a national park. The wolves had also stumbled onto a place without competing bears or cougars. They could easily run off the coyotes and coexist with the nimble foxes. As for prey, there were no white-tailed deer, but plenty of moose and beavers—with adventuresome or perhaps wayward individuals from both species making near-impossible swims and then homesteading there.

Since 1959, wildlife biologists, first from Purdue University and then from Michigan Technological University, had been crisscrossing the island during winter, mainly by ski plane, looking for moose and wolves. With the winter snows covering up autumn's fallen leaves, they had a bright white backdrop for their census work and a clearer view through the trees. The researchers also put boots on the ground, finding and inspecting the remains of moose carcasses, checking scat for clues on what the animals ate, and hunting for skulls and bones—all to piece together the dynamics of the island ecosystem. Wildlife ecologist Durward Allen started the project, Rolf Peterson took it over for the next four decades, and then John Vucetich joined him and took the reins within the past decade. Their collective works of observation and measurement have become the longest uninterrupted study of predator–prey relationships in the world. Isle Royale was a geographically isolated, stripped-down ecological laboratory dominated by wolves and moose.

I'd been following the ups and downs of the park's wolves ever since my original journey there, and with great anticipation, I returned in August 2015, when the late-summer sun had wrung the sharp cold from the water and warmed the breeze. This time, I arrived by seaplane with US Senator Gary Peters of Michigan, a leading voice on animal welfare issues, and Eric Clark, a wildlife biologist with the Sioux St. Marie Tribe of Chippewa Indians. John Vucetich met us in Houghton and flew over with us, and after we touched down, we took a quick boat ride to pick up Rolf Peterson. Peters, Clark, and I had two guides and interpreters who had covered nearly every inch of terrain on the island and knew every long-standing wolf and moose trail.

According to Peterson and Vucetich, it was a time of crisis. They counted just three wolves on the entire island in the previous winter—with the youngest of the group showing a curved spine and other genetic abnormalities. The situation was more desperate than at any point since Allen began his studies more than a half century earlier. When I worked here in 1985, there were three packs, with twenty-two wolves in all—and that was after a parvovirus epidemic had halved the population. A visitor, not adhering to Park Service policy, had brought an infected dog on a backpacking trip, and with no past exposure the wolves lacked immunity. The wolves' losses back then had nothing to do with lack of prey—there were one thousand or so moose.

Now, something else had decimated the wolf population of Isle Royale. "By 2009, we realized we had an inbreeding problem," Vucetich told me, with too few wolves, too closely related. The wildlife scientist explained that he and Peterson had discovered that the wolves had bone abnormalities and other physical defects that were indicators of a genetic bottleneck, making it harder for them to breed, and to kill big and powerful moose. Inbreeding had almost certainly cut down their resiliency. It had prevented a healthy bounce back from the parvovirus outbreak that hit in 1980, and now it was hindering them from even maintaining a viable population. The only thing that could improve their fortunes was an infusion of new genetic material from wolves not of this place—whether new immigrants crossing Lake Superior, or other animals relocated there by National Park Service personnel and other government wildlife managers.

The debate over the role that inbreeding played in the wolves' decline, Vucetich told us, had been put to rest when he and Peterson decoded the "very powerfully beneficial" effect of a single immigrant wolf who'd come across an ice bridge in 1997. Old Gray Guy, as they named him, performed an unintentional act of "genetic rescue." He impregnated a number of females, who birthed pups healthier and more robust than the ones born to parents long resident on the island. Old Gray Guy "made us realize that there was no way Isle Royale wolves could have persisted all these decades unless that phenomenon was happening with some regularity." Before then, he explained, "we figured it occurred once every fifty or sixty years." Once he and Peterson saw the effect of that single wolf, Peterson pored back over his decades of observations and realized that he'd missed some telltale signs of immigrant wolves who'd had similar effects on the health of the island's wolves. But with warmer winters making ice bridges less common, there were now fewer opportunities for wolves to cross Lake Superior.

The wolves—the central characters in the Isle Royale story for more than a half century and thus the primary draw for tourists—

were vanishing. Vucetich and Peterson had sounded the alarm as early as 2011 and called for intervention, but the National Park Service didn't seem inclined to respond. In 2012, there was an accident, also of human origin, that contributed to the downward spiral—in fact, it was a loss of 25 percent of the population. Two males, including a dominant one, and one female fell into a nineteenth century mineshaft that predated the origin of the park and drowned. At the time, there were few breeding females, and these mine-related deaths were more than a minor jolt to the declining population. Vucetich and Peterson and his wife found the bodies in the shaft, and pulled out the wolves' remains to confirm their fate.

Administrators took it all under advisement, citing the park's status as a wilderness area and its policy of letting nature take its course. Yet it was hardly an unthinkable proposition for the Park Service to grant permission to introduce native predators. Twenty years earlier, the Service had helped introduce Texas cougars into the Everglades ecosystem to genetically rescue the isolated population of endangered Florida panthers, weakened by inbreeding. The Park Service had also helped release Canadian wolves into Yellowstone in the mid-1990s to restore apex predators to a place where they'd lived for thousands of years and to check growing elk and bison populations. That occurred despite vehement objections from ranchers, hunters, and many other locals who feared wolves' impact on cattle, elk, and deer. But unlike Yellowstone, Isle Royale had no local opposition to wolves, since nobody lived there (the park is closed to visitors for the colder half of the year). As with Yellowstone and the Everglades, reintroducing predators on Isle Royale would be an attempt to undo a disruption caused by humans, who had delivered the deadly parvovirus, set a trap with an empty mine shaft, and warmed the atmosphere enough to limit migrants from Canada. Even a remote setting like this—closed to wolf killing and buffered from the effects of civilization by the world's largest lake—was not immune from human impacts. Helping the wolves get back on their feet meant unwinding the damage we had done, to say nothing of the importance of their role in checking the growth of an abundant moose population.

Letting the wolves languish and die off wouldn't just upset the entire Isle Royale ecosystem; it would also harm the local mainland economies. With the main actors in its ecological drama fading from the scene, and the moose population set up for a boom-and-bust cycle without wolves to check their growth, a whole lot of would-be park visitors would just go elsewhere. It would be like a play with no leading characters, and the entire moose-and-wolf storyline threatened to be a look-back rather than a present-day narrative. Visitor numbers might drop dramatically. Wolves shrink moose numbers to healthy levels on the island, but they swell the number of people who go there. There'd be less in the way of ticket buyers ferrying to the island, and the Park Service might cut some of its eighty full-time employees to compensate. The staging areas for Isle Royale tourists—Houghton and Copper Harbor in Michigan and Grand Portage in Minnesota—would feel the effects too. People come to the island for the hiking, scenery, and solitude, but they stay there for the wolves and moose. "They would love to see a print or hear a wolf, but if they don't, they are not disappointed," Vucetich told me. "Isle Royale is a great place to love them even if you don't see them." On one of our hikes, Vucetich pointed to a fresh wolf track in the mud, and the sight of it reminded me that even a handful of hard-to-find wolves make their presence felt.

The economic benefits that national parks—called America's "best idea" by nature writer Wallace Stegner—bring to nearby communities are immense. This is where the federal government exerts its power to drive the humane economy. Once designated as parks, these places—from Isle Royale to the Everglades to Big Bend—gain a "halo effect" and become top-tier destinations for tourists. They drive commerce in the regions abutting the parks, with hotels, shops, restaurants, coffee houses, camping outlets, guides, and other businesses springing up. In

2013, the National Park System received more than 273 million visitors at 401 land areas—the equivalent of almost every American citizen paying one visit. All those visitors spent $14.6 billion in surrounding gateway regions (defined as communities within sixty miles of a park), which helped create 238,000 jobs and contributed some $26.5 billion to the nation's economy. In creating a park, Congress provides a shovel-ready stimulus package—with no shovels needed. Mother Nature took care of the foreground, the backdrop, and everything in between ages ago, and takes care of the landscaping every year at no charge.

The workings of nature are not an economic activity, but just for perspective some studies have tried to put a price on it. According to authors Peter Diamandis and Steven Kotler, "The broader value of 'ecosystems services'—including things like crop pollination, carbon sequestration, climate regulation, water and air purification, nutrient dispersal and recycling, waste processing, and the like—'has been calculated at $36 trillion a year,' a figure roughly equal to the entire annual global economy." There are plenty of marvels of human ingenuity—big dams, voluminous digs for tunnels, buildings as tall as mountains, and alpine roads that defy imagination—yet as natural wonders the parks are more brilliant, soothing, and grander than just about anything humans could engineer. There is development around the perimeter of a park—hotels, shops, and concessions—but its scale and architecture are meant to align with the ethos of a nature experience.

Even today, when a citizen's group or political leaders call for designating a land area as a park or a national wildlife refuge, they typically describe it as an economic development opportunity, given the spending of tourists that inevitably follows. That's exactly the argument of the boosters for the proposed Maine Woods National Park, which would revitalize the economies of struggling mill towns like East Millinocket and Medway that today have few good-paying jobs and scant incentives for young people to stay. Far from locking up land, as critics of public land designations claim, such action amounts to unlocking its greater potential for broad, lasting benefits. What started as an American idea is now a global one. In a nation whose people possess an extraordinary work ethic and stress levels, nature is the rage in South Korea; according to *National Geographic* magazine, "visitors to Korea's forests increased from 9.4 million in 2010 to 12.8 in 2013." And unlike extracting oil or natural gas or minerals, which are finite resources, these places provide an endless supply of what people want—the overlooks and vistas, the forests and rock formations, waterfalls and streams, and the birds and mammals that people never tire of seeing from a new angle or perspective. There is increasing evidence of the emotional and physiological benefits of seeing wildlife and spending time in nature, and that delivers incalculable economic benefits.

Something deep within us explains the desire and need to visit national parks and other natural areas. For many of us, the experience offers a kind of peace that can be found nowhere else. Maybe it stirs a sense of homecoming too, a bit like returning to the scenes of our childhood. For 99 percent of the human story, after all, tribal societies lived amidst animals, trees, streams, and all of the other features of nature. The Harvard biologist E.O. Wilson calls it "biophilia"—our love of nature, our "innate tendency to focus on life and lifelike processes." Today, escaping from the bustle of our office complexes and subdivisions and immersing ourselves in unspoiled areas allows us to exhale. We're drawn to these places, as Wilson says, "like a moth to a flame."

During the last 150 years, apace with society's industrialization, urbanization, and inexorable development, we've also acted to save natural areas. More than a century ago, urban planners created green spaces in our biggest and best-known cities. Today, vast numbers of New Yorkers stroll through Central Park, and Washingtonians breeze past streams and forests in Rock Creek Park. In fact, for homes that overlook or abut the parks, sellers can add tens of thousands of dollars to the list price—tangible evidence that the value of nature can be monetized. In addition to city parks departments, there are multiple state and federal agencies, private conservancies and land trusts, and even private landowners who protect thousands of other tracts of land and forest. The Park Service just happens to be the chief custodian of a collection of the natural world's most special reserves, but nature's bounty has countless stewards. We moved on long ago from living in untrammeled places, but we still need those places to live.

Saving nature and allowing human access and controlled uses has been a bulwark of our economy for more than a century. During the Great Depression in the 1930s, President Franklin Roosevelt and Congress created the Civilian Conservation Corps to give young men jobs building trails and access through natural areas for public enjoyment. His fifth cousin and predecessor, Theodore Roosevelt, believed that preserving and experiencing open space was critical to character development and referred to the "strenuous life" as a unique part of the American experience. Though he was almost pathological in killing North American and African wildlife for trophies and authorizing the US Bureau of Biological Survey to rid the West of wolves, the first President Roosevelt will always deserve credit for setting aside a vast network of places for public protection. During his nearly eight years in office, he created 150 national forests, 51 federal bird reservations, 18 national monuments, 4 national game preserves, and 5 national parks. Altogether, Theodore Roosevelt placed 230 million acres of land under permanent public protection—the equivalent of the landmass of the Atlantic coast states from Maine to Florida. Nature preservation has been an enduring feature of the American economy and an essential part of our culture ever since.

Behind the creation of every park were a handful of advocates who had a vision for saving a beautiful place, and who stayed the course in the face of initial indifference and opposition by special interests. The successful campaigners eventually assembled coalitions that persuaded lawmakers and presidents to take action. They were social entrepreneurs of a different sort—shielding rather than building, preserving rather than exploiting. Their labors made the politicians who authorized these parks look provident, since other uses of the land would never have generated so much commerce, and done so as sustainably, humanely, and lastingly.

The result has been an eco-commercial boom of unprecedented scope and impact. In 2013, Yellowstone Park alone received 3.2 million visits, worth $382 million, creating 5,300 jobs in gateway communities. (The Greater Yellowstone ecosystem, which also encompasses Grant Teton National Park and an array of national forests and national wildlife refuges, generates a billion a year in nature-based tourism revenues.) Yosemite received 3.7 million visits, worth $373 million, creating 5,033 jobs. Great Smoky Mountain National Park in eastern Tennessee received more than ten million visits, even though many of them pass through quickly. By comparison, the fifteen-to-twenty-thousand people who go to Isle Royale seem a relative trickle, but they often stay for a week at a time—more than twice as long as the average visit to Yosemite. In these rural economies, this infusion of visitors and their loading up on supplies for a long stay pays the bills.

The Park Service must manage the flow of people and shield the parks from the effects of too much love and appreciation, but it also must build understanding of the parks' science and ecology. That's where John Vucetich and other scientists come into play—as the most highly trained and insightful of chroniclers, documenting the biological dynamism of these places. Durward Allen and Rolf Peterson did that for me when I was a kid, and now Peterson and Vucetich are continuing this work for the next generation. Other scientists recount the lives of grizzly bears in Yellowstone, of panthers in the Everglades, and of Key deer on the southern tip of Florida. They are joined in this task by the filmmakers, photographers, publishers, essayists, and reporters who also tell parts of the story, delivering details to mass audiences that excite further interest in these places. Together, they are the marketers of the humane economy. They remind us that animals make the landscape come alive, stir our hearts, and cause our eyes to widen in awe. The humane economy requires an ensemble cast—and that's never more evident than ever in the workings of national parks, with government, scientists, businesses, and consumers playing key roles in this economic ecosystem.

The scientists are also there to help identify threats. Senator Peters of Michigan, our companion on the August 2015 trip to Isle Royale, has been critical of the Park Service's hedging over the dissolution of the island's wolf packs. Months earlier, the senator called for the Park Service to conduct a genetic rescue, noting that the "three remaining wolves may struggle to reproduce, and if they do produce offspring, the tiny genetic pool will lead to inbreeding and further complications." He had enlisted his Michigan colleague, Senator Debbie Stabenow, and other key lawmakers in supporting more action. "Unless the NPS acts quickly," they wrote, "wolves are almost sure to disappear from Isle Royale," and that in turn would "lead to significant, harmful changes to the ecosystem in this remote park." Before wolves arrived in the 1940s, Isle Royale's moose population grew so large that the animals denuded the forest of much of the edible vegetation.

The plan to trap wolves and relocate them to Isle Royale is an elegant solution for an additional reason. One nearby reservoir of wolves is the Upper Peninsula of Michigan. Taking some wolves from there would have the effect of relieving angst among some residents about their presence there, where there are vast forests but also human settlements. During the same November 2014 election in which Michigan voters elected Gary Peters to the US Senate, they rejected two statewide referendums to open up trophy hunting and trapping seasons for wolves in the Upper Peninsula. Commanding majorities of Michiganders sided with the wolves.

Yet sentiment to kill wolves remained intense among some in the Upper Peninsula, where the culture more closely resembles rural parts of Alaska than the urban enclaves of Detroit or Grand Rapids. Wolves had reclaimed these northern woods during the last two decades, emigrating from Wisconsin, and, according to Michigan Department of Natural Resources estimates, there were more than one hundred packs surviving there in 2015. With 300,000 people in the Upper Peninsula and just more than 600 wolves, they were bound to cross paths occasionally, even in the vast, thinly populated expanses of a peninsula that is 320 miles east-to-west and that comprises nearly a third of the land mass of Michigan but is home to only 3 percent of its human population. Some of the inevitable and recurring conflicts could be solved by removing any offending wolves and sending them to Isle Royale.

This wouldn't be a long journey for captured wolves, and with no permanent human residents or farming in the national park, it would be difficult for them to get into trouble. In this case, being exiled to an island wouldn't be all that bad either. They'd be free of any interaction with people for half the year, and Vucetich said that even wolves who'd been accustomed to hunting deer on the mainland could adapt and make moose their prey. The Detroit Zoo (which also offered to take a few wolves) and the HSUS agreed to pay for the costs of capture and transport for wolves who'd been caught menacing livestock. If translocation succeeded, there would be benefits aplenty—for the park in restoring its famed wolves, for the businesses needing a draw to keep the visitors coming, and for those in the Upper Peninsula who feared wolves and wanted the problem ones gone.

The plan could also ease some of the tensions associated with the battle over the 2014 referendums in Michigan. In their campaign to authorize a wolf hunt, proponents had been unsparing and apocalyptic in branding wolves as terrifying marauders. They trotted out the experience of an Upper Peninsula farmer, John Koski—operating in the far western portion of Michigan—who claimed that he'd had countless visits from wolves and had dozens of livestock animals killed. Koski's farm was the site of more than 60 percent of all wolf attacks on livestock in Michigan. Yet reporter John Barnes, working for a consortium of Michigan newspapers under the banner of *M-Live*, exposed the whole narrative spun by hunt proponents as a trumped-up argument for hunting and trapping. It turned out, according to a months-long investigation by Barnes, that Koski (whether inadvertently or not) had been baiting wolves with deer and cow parts and then bellyaching about wolf incidents—in addition to getting financial compensation for it. The state even gave him some guard donkeys (it turns out donkeys are very territorial and will alert and seek to protect their fellow livestock from intruders), but Koski failed to use them properly—two of them actually died on his watch and officials had to intervene to save the third one. The reporter found that the Michigan Department of Natural Resources, which has a history of closely serving the hunting lobby, was also using the details of Koski's experience as a rationale for its plans to carry out the hunt. He concluded that "government half-truths, falsehoods, and livestock numbers skewed by a single farmer distorted some arguments for the inaugural hunt," which took place in 2012.

Some months after John Barnes's *M-Live* series ran in papers throughout Michigan, Koski pled no contest to charges of animal neglect—further undercutting the whole shaky argument that had been advanced to justify the hunt in the first place. The lid on hunting, at least in the short run, was closed even tighter after a US District Court judge ruled, just a month after the votes by Michigan citizens against wolf hunting and trapping, that the federal government's action to remove protection from wolves had been unwarranted and illegal. The court restored federal protection for wolves throughout the Great Lakes Region, casting doubt on future hunts in Minnesota, Wisconsin, and Michigan. (A separate suit shut down Wyoming's hunt in 2014.)

Politicians in the Upper Peninsula had been clamoring for the Michigan hunt for quite a while, since around the time wolves started reclaiming land their ancestors had roamed for thousands of years. The US representative whose district includes the entire peninsula, Republican Dan Benishek, introduced legislation to overturn the court ruling that restored federal Endangered Species Act protections for wolves. He was, in turn, recycling the arguments of Koski and state politicians who claimed that wolves were regularly attacking livestock and putting people in danger. But Benishek was a more restrained voice than State Senator Tom Casperson, the chief anti-wolf crusader in the Michigan state legislature. Casperson got in trouble for exaggerating the case against wolves. He concocted a menacing story out of a rather benign incident where a constituent of his spotted a wolf on her lawn in 2010. Casperson highlighted this story in introducing a resolution to restore the state's ability to open hunting and trapping seasons and to remove federal protections: "Wolves appeared multiple times in the backyard of a daycare center shortly after the children were allowed outside to play," his resolution read. "Federal agents disposed of three wolves in that backyard because of the potential danger to children."

John Barnes looked into this story too. He reported that "there were no children in the backyard that day. There was a single wolf, not three. Three wolves eventually were shot—seven months later, on three separate days, on a plot roughly three-quarters of a mile from [the constituent's] property."

Tom Casperson took enough heat over his statement that he was forced to apologize on the Senate floor for his scaremongering, but that incident didn't give him pause in his efforts to promote the killing of wolves. "We are concerned for our children's safety in our backyards," he declared in defense of his plan to open up a sport hunting and commercial trapping season. He neglected to mention that of America's big predators, wolves have proved among the least menacing—with no documented fatal attacks by healthy wild wolves on people in the lower forty-eight states in the last century. That's right, none. It's true, scientists tell us, that the amygdala in our brains is hard-wired to fear predators, even in cases, as with wolves, where the data suggest that the risk of a dangerous encounter is essentially nil. But when it comes to establishing policy, hard data should outweigh vague fears—and it did when Michigan voters went to the polls in November 2014 and rejected Casperson's arguments, restoring protections for wolves.

Taking an even longer view of the human–wolf relationship, let's remember that this species came to us in friendship perhaps more than thirty thousand years ago, according to an emerging consensus among paleontologists. Pre-agricultural people and wolves may have viewed each other warily at first, but eventually an instinctive fear turned into acceptance and then later into companionship. Of the twenty or so wild animals that humans have successfully domesticated through the ages, wolves were first, by twenty thousand years or more—and the wolf is the only large predator ever successfully domesticated. Though the other large predators, whether mountain lions or black bears, generally don't pose much of a threat to us either, the wolf is the only one who became a companion. And while the wolf has survived through the millennia, in a dramatically reduced range as a result of human persecution, it has also offered an extraordinary array of benefits to human society by submitting to domestication. At the top of every dog's genealogy chart is the wolf, and dogs have provided us with companionship, security, labor, and enhanced success in agriculture (guarding) and hunting (tracking and retrieving) over the millennia. More than any other species, the wolf and the dog enabled us to make the leap from tribal societies and prehistory to agriculture and the rise of human civilization. "Dogs absolutely turned the tables," Dr. Greger Larson, director of the Palaeogenomics and Bio-Archaeology Research Network at the University of Oxford told the BBC. "Without dogs, humans would still be hunter-gatherers. Without that initial starting phase of dog domestication, civilization just would not have been possible."

The story of a wolf suckling the abandoned brothers Romulus and Remus at the creation of Rome may be allegorical, but wolves did forever change the human story. Their domestication stands alongside language, fire, and plant cultivation as one of the major innovations that most altered the fortunes of humanity. "It's hard to see how early herders would have moved and protected and guarded their folks without domestic dogs being in place, and one has to wonder whether agriculture would ever have really made it as a viable alternative to hunting and gathering," said Peter Rowley-Conwy, a professor of archaeology at Durham University. It's no exaggeration to say that nearly all human exchange, from early bartering to the online transactions in the information age, have necessarily been built on the foundation stones of these developments. The wolf and its descendants have been part of the humane economy longer than any other species.

While many Native American tribes revere wolves, and place them at the center of their creation stories, many other Americans have succumbed to the cartoonish "big bad wolf" narrative and done the opposite. Throughout much of US history, wolves have been ruthlessly persecuted. In his 1880 annual report, Yellowstone National Park Superintendent Philetus Norris wrote that "the value of their [wolves and coyotes] hides and their easy slaughter with strychnine-poisoned carcasses have nearly led to their extermination." Around the turn of the century, the federal government hired professional hunters and trappers to amass a body count of wolves, and state governments provided bounties on them. By the early 1970s, just after the US Congress enacted a comprehensive Endangered Species Act to protect them and so many other imperiled species, wolves were hanging on only in the northern reaches of Minnesota—and at Isle Royale in Lake Superior.

Increasingly wildlife science is revealing the critical part that wolves play in maintaining robust ecosystems. Aldo Leopold, the father of modern-day wildlife management, renounced his killing of wolves in his classic book, *A Sand County Almanac.* He came to recognize, even in his days as a young forester, that wolves were anything but pests. They were critical actors in maintaining balance in ecosystems, and he saw the harmful effects of their removal, by predator control programs, in Arizona's Kaibab National Forest. In their decades of work at Isle Royale, Vucetich and Peterson have affirmed Leopold's conclusion beyond all argument by showing how wolves limit the growth of prey populations—strengthening them by culling the weak, sick, or young, and preventing their numbers from expanding to the point where they denude the forest of saplings or strip bare the leaves of trees. Indeed, upon their reintroduction to Yellowstone, wolves immediately went to work reducing the high densities of elk and bison, forcing them to stop overgrazing meadows and riparian areas. These effects are documented in a popular video called "How Wolves Change Rivers," based on a lecture by journalist and environmental advocate George Monbiot. The video has attracted more than fifteen million views on YouTube.

The effects of wolves on livestock are also overblown. Data from Michigan's Upper Peninsula and other parts of the country where wolves live show that they are responsible for a very small amount of killing—between 0.1 and 0.6 percent of all livestock deaths in these areas. A 2014 Washington State University study, conducted over a twenty-five-year period, found that indiscriminate killing of wolves actually increases the tendency of wolves to prey on livestock. The reason may be that sport hunting and commercial trapping of wolves break up stable wolf packs, creating a younger, less experienced population, inexperienced in killing traditional prey and more likely to show opportunism and pick off a sheep or calf. And of course, farmers who deploy guard dogs as a highly successful strategy of protecting their flocks and herds from predators can thank the wolf itself for that service.

In exaggerating the adverse impacts of wolves, the proponents of wolf killing underreport these beneficial effects. Wolf predation helps maintain healthy deer populations, to the benefit of forestry, agriculture, and wildlife management. By killing sick deer, wolves can contain the spread of diseases that can be catastrophic for deer populations. And what automobile drivers haven't been concerned, to one degree or another, by the possibility of colliding with a deer on the road? The insurer State Farm reports that there are roughly 1.2 million deer–vehicle collisions in the United States every year, causing some two hundred human fatalities and about $4 billion in vehicle damage. Michigan typically accounts for about fifty thousand of those collisions a year, with thousands of them in the Upper Peninsula. If the wolves maintained viable, healthy herds by taking mainly the young, weak, and sick deer, they might save tens of millions of dollars in repair and insurance costs, to say nothing of the incalculable benefits of preventing the loss of human life.

"Wolves provide a firewall against new diseases in deer," Rolf Peterson told a Michigan Senate committee when the issue of wolf hunting was being debated. "A very obvious example may be chronic wasting disease." Chronic wasting disease (CWD) is a brain disease, like Mad Cow Disease, and it's one of the major threats to deer populations, after it spread from deer farms and captive hunting facilities to free-ranging deer populations—another case of reckless trophy hunters visiting more affliction upon wildlife. The Wisconsin Department of Natural Resources, which says there are more than four hundred deer farms in the state, notes that the disease has been spreading in large portions of the southern part of the state, since a major outbreak of the brain disorder in 2002. "So far CWD has not spread into areas inhabited by wolves, anywhere in the United States," Peterson said, "and the logical hypothesis is that wolves simply cull out diseased animals." The disease is an ugly and nonselective way of reducing deer populations, and it creates a health risk to people who eat deer meat since the disease can be transmitted to people through its human variant, Creutzfeldt-Jacob Disease.

And while the indirect economic benefits that wolves bring may provide their greatest value, there are direct benefits, too. As with Isle Royale, people also just love to see wolves; to hear wolves; and to place themselves, even for a short while, in a wild place that harbors wolves. Each year, thousands of wildlife watchers gaze at the world's most-viewed wolves in the Lamar Valley of Yellowstone, bringing in $35 million to the Yellowstone region annually. In the Great Lakes region, the International Wolf Center in Ely, Minnesota, receives $3 million each year from wolf watchers. With wolves now claiming a more permanent place in the Great Lakes Region, we can expect tourism-related revenues to increase in all of the states with wolves.

Vucetich is leery of invoking the practical, economic arguments about wolves, though he readily acknowledges their validity. It's just that he believes the moral argument for protecting wolves is the most important and compelling. He understands, however, that political decisions and public policy more often turn on economics. Ethics and economics are bound together in our decision making, and it's clear that many politicians have it upside down when it comes to the economic analysis associated with predators. It's not a zero-sum game, where more wolves mean fewer hunting licenses and increased cattle and sheep losses. A more comprehensive, fact-based assessment of wolves shows their multiplier effect—whether in a pure wilderness like Isle Royale or in a landscape where forests, farms, and human settlements are commingled. "I think there are several areas where wolves are already playing a role in remaking the Upper Michigan ecosystem," Peterson told lawmakers in his testimony. "The total positive economic impact would be measured in hundreds of millions of dollars."

For more than a century, we as a nation had our way with wolves, killing them off throughout more than 95 percent of their historic range—leaving a trail of broken bodies and shattered wolf families and a gaping hole in our ecosystems. Like so many other forms of animal abuse, much of the killing was driven by ignorance and misguided government action. That decades-long scorched-earth policy of slaughtering wolves stands alongside the massacre of bison as one of the most inhumane and counterproductive chapters in the annals of American wildlife management and agriculture. For too long, government policies toward wolves were driven by fairy tales and irrational fears. When we see people reverting back to such arguments, dusting them off for use on the floors of state legislatures or in meetings of local cattlemen's associations, it's time for us to call them as they are—false, groundless, and shameful. We know too much now to let claims like these carry the day any longer.

We cannot be Pollyannaish when it comes to the occasional conflicts that arise with wolves, but we need not be foolish or cruel in our responses either. We must close the door on the era of indiscriminate killing and deal with occasional conflicts primarily by nonlethal means, as part of a multidimensional framework for management that takes reason and predator ecology into account. In recent decades, we've tried to undo some of the damage: first, in the 1970s by protecting the small population of surviving wolves in Minnesota and Michigan through the Endangered Species Act, then two decades ago by reintroducing wolves to the Northern Rockies and the Southwest. These toeholds have allowed wolves to reclaim lands lost to them generations ago. The wolves have demonstrated enormous resiliency, but there's no need to continue to test that capacity. They've reclaimed forests in Michigan, Minnesota, and Wisconsin; established packs in southern Arizona and New Mexico; and now in northern California and throughout much of Oregon and Washington. They've even wandered into northern Arizona and into Utah. Yet for all this progress, only five thousand or so wolves survive in the lower forty-eight states.

In the last few years, the US Fish and Wildlife Service removed federal protections under the Endangered Species Act for wolves in a number of states and turned over management to wildlife authorities there. Without hesitating, these states authorized trophy hunters and trappers to kill thousands of wolves, often with steel-jaw traps, wire snares, hounding, and baiting. In Wisconsin, where the worst of the methods were permitted, trappers and trophy hunters killed off seventeen family units in just three seasons, or a fifth of the state's total wolf population.

The courts have provided a check on recent abuses, repeatedly stepping in to restore federal protections for wolves in Wyoming and the Great Lakes. Congress overruled the courts in the Northern Rockies, and Montana and Idaho treated that act as a license to kill wolves in appalling numbers. Thus far, HSUS and others have blocked similar efforts in Congress to remove federal protections for wolves in the Great Lakes and Wyoming. It's the wrong move not just as a matter of law, but also as a matter of ecology and economics. In late 2015, dozens of world-renowned wildlife biologists and scientists wrote to Congress, noting that "the gray wolf occupies a mere fraction of its historic range." "In recognition of the ecological benefits wolves bring, millions of tourism dollars to local economies, and abundant knowledge from scientific study," they wrote, "we ask Congress to act to conserve the species for future generations."

Having played a central role in protecting our parks and generating billions of dollars in tourism, all while preserving the cultural and ecological assets of our nation, government should augment its role as an agent of the humane economy, and the protection of wolves, grizzly bears, wolverines, lynx, and other long-persecuted creatures presents the perfect opportunity. We now know too much about wildlife science to continue our old ways, and we have too much evidence of economic benefits to be in denial any longer. If ecology and economics are not enough, we need only look to the occupants of the dog beds in our homes, or the ones who sleep in our beds. They're not so fierce, but they're dependent on us. We can be good caretakers of them and their wild brethren and other wild animals, who need little more from us than to stop killing them for sport, bragging rights, or some irrational hatred.

Family Planning and the Future of Wildlife Management

Ms. Manners might not approve, but the rest of us were fine with what Jerome Fox had up his sleeve—in fact, way up his sleeve. He had his arms pinned against his sides, and he'd tucked a frozen vial of a horse anti-fertility drug under each armpit. Sweat beads dappled his faded shirt. His cap—emblazoned with the words WILD HORSE PROGRAM—shielded his face from a July sun so potent that it vaporized every wisp of cloud cover. The sixty-one-year-old wild horse specialist with the Bureau of Land Management (BLM) was hastening the thawing process with a highly strategic use of body heat. He was prepping the porcine zona pellucida (PZP) immunocontraception vaccine for use in a dart gun that looked menacing enough to be in the clutches of a US Marine. Fox's plan was to direct the dart into the rump of a young mare before she and her twenty herd mates about a hundred yards away got wise to him and scattered, with their signature tails and manes trailing them in the wind. It was family planning out on the open range—the US version of a one-child policy but for wild horses.

Out in the expanses of the Sand Wash Basin Herd Management Area—a short dash for wild horses from Colorado to the Utah and Wyoming borders—there's not a house or human resident to be seen. Wild horses are long lived, and with the right conditions, they can double in number over a five-year period. On these 150,000 acres of federal land, there are no wolves about, and mountain lions and black bears aren't much of a threat, because the horses are typically too strong and fast for them. The horses' biggest competitors are cattle and sheep and also deer, elk, and pronghorn, but they are just proxies for the ranchers and hunters who have an economic or recreational stake in hoarding forage for their designs. In the Sand Wash Basin, the BLM gives three grazing allotments—two for ranchers running sheep and one for cattle. It's largely because of public lands grazing, and the BLM's edict to manage for "multiple-use" and its obeisance to the ranching lobby, that there's such a big to-do over the presence of wild horses. It's not enough that the ranchers get a sweetheart deal and pay below-market grazing fees—they want the BLM to remove the wild horses too. For decades, BLM has been their willing partner.

"They are very tolerant because they are so photographed, and that means that shooters can sneak up close enough to put a dart in their rump with a gun within a forty-yard range," said Stella Trueblood, a lead volunteer in the fertility control program and part of the team of five that day that included my HSUS colleagues Stephanie Boyles Griffin and Kayla Grams, the BLM's Fox, and me. With a rancher's tan, and a long, graying ponytail to rival that of any mare, Trueblood could pass for a venerable descendant of the Cheyenne Indians who once called this land their own. But the sixty-year-old is actually a native of Britain and acquired the Trueblood name in marriage. She'd come to know most of the horses by the brilliance of their colors and markings—dappled gray and brilliant red duns, along with bright chestnuts, buckskins, paints, and palominos.

The BLM began partnering with the HSUS to initiate fertility control in Sand Wash Basin in 2008, rounding up sixty-two mares and then hand-injecting them with PZP before releasing them back onto the open range. (One group was given a one-year formulation of the vaccine, while another group was given a longer-acting one, to test the relative efficacy of the vaccines.) In treating penned horses, there's no need for sharpshooting with a dart gun—just a jab stick with a needle on the end. But that process is not only expensive; it's extremely stressful for horses who've never been touched by a human hand and are often chased by helicopters for miles across rugged terrain before they're driven into corral traps.

Kayla Grams, our HSUS wild horse specialist, hit upon a new solution. She worked with seasonal wildlife technicians to begin the remote, opportunistic darting of fifty-one of the previously treated mares with PZP boosters. In 2013, Grams, Fox, and then-HSUS staffer Josh Irving managed to dart 118 mares—including 36 previously treated and 82 untreated mares—in the Sand Wash Basin. Trueblood tagged along with Grams and Irving to pick up on their darting techniques. After that, in 2014, Trueblood and Fox took the lead in darting the Sand Wash horses, working to get PZP into any remaining, untreated mares, and then hitting them a second time with a booster. "They do very well out here even though it is so dry," Trueblood told me. "Their survival rates constantly surprise me."

The vaccine, mixed with an adjuvant that triggers and intensifies the horse's immune system response, stimulates a horse's antibodies, which then attach to the sperm-receptor sites on her unfertilized ovum. The antibodies hinder sperm from binding to the ovum at these receptor sites, blocking fertilization and preventing pregnancy. Prior to 2013, PZP was classed as an experimental drug, needing authorization for specific uses by the Food and Drug Administration. But now it is registered by the Environmental Protection Agency under the brand name ZonaStat-H and has been registered for use in more than ten states. For horses, the first treatment they receive in their lifetime is referred to as a "primer" dose of PZP; a year or more later, the mares receive a booster dose. Grams oversees the HSUS's involvement with the contraception program at Sand Wash and with similar programs for the Cedar Mountain herd in Utah and Jarita Mesa herd in New Mexico.

"Last year, we had 104 foals born, and 96 or 97 survived," Fox said of the Sand Wash Basin herd. "With the help of HSUS, Stella and her crew and myself, we got in the neighborhood of 130 mares darted prior to the 2015 mating season. We have 57 born this year. We lost seven, so we are fewer than 50 live foals this year."

Throughout history, humans have done family planning for animals, mostly with a view to our own interests. We domesticated a dozen-and-a-half wild animals through the ages, and we've been selectively breeding them ever since—morphing wolves into Chihuahuas and Great Danes and every kind of domesticated canid in between, and creating hundreds of breeds of cats, horses, cattle, pigs, chickens, and other animals. We've controlled breeding of these animals by keeping males apart from females or by castrating the males, and, for dogs and cats, spaying the females. With PZP and other fertility control drugs, including GonaCon, we're taking it all to another stage in which compounds and chemicals will do the trick without the need for surgery. What the pill did for humanity, introducing a measure of control and rationality to reproduction, PZP and related compounds can do for animals, which in their case can spare entire species a lot of grief.

Dr. Jay Kirkpatrick, the father of the vaccine and founder of the Science and Conservation Center in Montana, had treated free-roaming horses with PZP on Maryland's Assateague Island National Seashore for two decades, effectively capping the number of clattering hooves on the fragile beach grasses of this barrier island. Kirkpatrick, who passed away suddenly at the end of 2015, had also pioneered humane population management programs for elephants in South Africa—where the animals have achieved high densities in tightly bound protected areas, such as Kruger National Park. Traditionally the government handled high densities of elephants, along with the conflicts they create, with a practice euphemistically called "culling." In plain language, it was mass slaughter, with hundreds of elephants frantically running from aerial shooters, calves witnessing the death of their mothers before being executed themselves, and fields left strewn for miles with the bodies of one of Earth's most glorious inhabitants.

Even the most hard-bitten hunters enlisted in these killing frenzies could be brought to tears by the sight of it, and it would be hard to picture any scene more in need of the new ways of the humane economy. Kirkpatrick and our Humane Society International team worked with the South African National Parks to put fertility control plans to work in more than a dozen parks and private reserves. Now with a track record of success, there's been broad acceptance of PZP, and it's trumpeted the end of the era of culling elephants in South Africa. It's a better outcome for all stakeholders, whether the wildlife managers, farmers, villagers, and, most of all, these wonderful creatures themselves.

About four months before I trekked to Colorado to see the reproductive lasso thrown over wild horses, I'd been with Stephanie Boyles Griffin and Josh Irving on a humane hunt for white-tailed deer in Hastings-on-Hudson, a bedroom community of eight thousand with tightly packed homes an hour north of Manhattan. "It is a practical solution in a densely populated community, where lethal options are problematic," Mayor Peter Swiderski told me as we sat down at a Main Street diner. And Boyles Griffin added that the goal is, from year to year, to treat thirty to fifty of the female deer. "Our goal in Hastings is to treat at least two-thirds of the female segment of the population," she said, "in order to curb the size of the population over time."

There are hundreds of communities where homes mingle with patches of forest that make edge habitat for deer, and where people are growing more impatient with the animals, especially when they dash across the road or eat ornamental shrubbery. If we can make fertility control work at Hastings, it has the potential to upend the conventional, lethal response to deer conflicts around the United States. Indeed, a majority of all US residents today live in the suburbs, and the deer have been flourishing in these areas right alongside them, so it's an issue on the minds of tens of millions of Americans. It's the rare person who, given a choice between killing off their wild neighbors or controlling deer populations, wouldn't think it's an easy call on the side of nonlethal methods.

Along with deer, perhaps the most intense demand for fertility control has been for horses. According to the BLM, there are about sixty thousand free-roaming wild horses and burros in the West, divided among dozens of herd management areas across ten western states, with the largest numbers in the sagebrush-dominated, high desert areas of Nevada and Wyoming. The federal Wild and Free-Roaming Horses and Burros Act of 1971 forbids shooting equids or rounding them up and slaughtering them, and assigns the BLM to manage and protect the animals. This law was enacted after a campaign by Nevadan Velma Johnston (a.k.a. "Wild Horse Annie") and millions of school kids, who demanded protections for these "living symbols of the American West." Johnston started this historic campaign after, one morning on her way to work in Reno, she saw blood trickling out of the back of a horse transport trailer—probably loaded full of wild horses by "Mustangers" who were known for their ruthless methods of rounding up and handling horses.

Without people shooting or slaughtering the horses by the thousands, as happened prior to the enactment of the 1971 law, they increased in number. In response, the BLM began rounding up the horses and removing them from public lands and adopting them to private citizens. In the forty-five years since the law was enacted, the government has rounded up more than 300,000 wild horses and adopted out more than 200,000—spending hundreds of millions of dollars in the process. Through the years, there have been plenty of documented cases of "killer buyers" masquerading as "adopters" caught red-handed shuttling the horses to slaughter plants, where they were butchered and shipped off to Europe and Asia for human consumption. Exposés in the press and scrutiny by advocates and Congress produced new rules that limited the number of horses any person could adopt and established penalties for trafficking in horses for slaughter. That didn't entirely stop killer buyers from getting their hands on rounded-up wild horses, but it slowed their illegal activity. In the end, some measure of fidelity to the law meant that the adoptions had to be even more brisk to prevent the swelling of captive wild horse populations.

The BLM, in the late years of the George W. Bush administration and the first term of Barack Obama, had been rounding up as many as ten thousand horses a year, spending a boatload of money to keep the free-roaming population at around thirty-five thousand. When the global recession hit in 2008, however, wild horse adoptions tanked. It costs plenty to feed, train, and stable a horse, and with so many people watching their stocks and 401(k) accounts contracting, they pared down expenses—and wild horses didn't make the cut. Despite that pullback in adoptions, the BLM kept up with large-scale roundups and removals and quickly had thousands more wild horses on its hands.

The department had to dramatically expand its leasing of long-term holding pastures, in the Midwest and West, to hold the newly captive horses indefinitely.

By 2015, the number of captive wild horses had reached 47,000—a population boom of sorts never imagined by anyone in the wild horse program. In all practical terms, the agency was now responsible for more than 100,000 horses, between the wild ones and the captives, and much of the problem was of the BLM's own making, with its unyielding roundups and its stubborn reluctance to implement fertility control, even after that approach had been proven to work on Maryland's Assateague Island. In a 2013 report called for by the Congress, a panel of the National Academy of Sciences determined that the BLM's removal of wild horses was actually contributing to population growth by increasing reproduction and survivorship on the range. "Regularly removing horses holds population levels below food-limited carrying capacity" and, according to the Academy report, triggers "growth from decreased competition for forage."

The Academy report concluded that the BLM had mismanaged its charge. "It is clear that the status quo of continually removing free-ranging horses and then maintaining them in long-term holding facilities, with no foreseeable end in sight, is both economically unsustainable and discordant with public expectations." None of this should have been lost on the BLM, since the agency doesn't pay to feed horses on the open range, but does pay for hay, real estate, and staff to oversee the captive animals. That cost is nearly $1,000 per horse per year, and the cost of the captive management program has exceeded $43 million a year—about two-thirds of the budget for the agency's entire wild horse management program. The cost to the BLM for keeping a relatively young horse in one of its holding facilities for life is $50,000, and that expense, multiplied by tens of thousands of animals, has amounted to a run on the bank. Finally, in 2013 and 2014, the agency realized that it could no longer stockpile more horses in holding facilities and ratcheted down the roundups to between three and four thousand horses per year—about the number it anticipated it could adopt. This has, in turn, resulted in the free-roaming wild horse population growing by perhaps five thousand or more a year according to the BLM, creating a problem requiring an innovative solution.

"Investing in science-based management approaches would not solve the problem instantly," the Academy report noted, "but it could lead the Wild Horse and Burro Program to a more financially sustainable path that manages healthy horses and burros with greater public confidence." With each passing year, it has become more evident that fertility control is the only answer. The Academy continued: "The porcine zona pellucida (PZP) vaccine, an immunocontraceptive, is the most extensively tested method in free-ranging horses and may be the most promising option at present."

In the Sand Wash Basin, we'd been driving for more than twenty-five minutes across the treeless landscape and had caught more than a few glimpses of prairie dogs and rabbits, and even ground-dwelling burrowing owls, before Fox and Trueblood spotted the band of two dozen horses, most of them with heads down and pushing aside sagebrush and saltbush to find a few shoots of grass. We pulled over, and Trueblood stepped out and gave the herd a careful once-over with binoculars before Fox started thawing the vaccine and mixing his antifertility cocktail. Trueblood said that all but two of the females had been treated, so those mares would be the immediate targets.

After Fox, a native of Roseburg, Oregon, and a lifelong elk and deer hunter, loaded the vaccine in the gun, I walked out with him and Trueblood, both of them armed with their dart guns and all three of us speaking in hushed tones. "I want that paint horse," said Fox. As we zigzagged toward the group, the lead stallion fixed his gaze on us, while most of the others seemed unfazed and continued foraging. "Out this far, you want them to see what you are going to do," he whispered. "They see me do all of this, so when I get closer, they say,

'Oh, I've seen this before.' " There was a small pop, more like an air rifle than the launch of a round from a high-powered rifle, and then a flash when it hit the mare. The dart dropped out immediately, but Fox said the vaccine, propelled by a small, controlled explosion in the dart, made it into her bloodstream. The herd, barely startled by the shooting, pushed ahead just ten yards before resuming their normal routine.

Trueblood had her eyes on a Palomino just about forty yards away. She took five steps toward the mare, used a monopod to steady her weapon, lined up the shot, and then fired, hitting her mark. Though she'd never handled a gun prior to joining the SWAT team, she'd turned out to be something of an Annie Oakley. Two for two they were. And it took no harried chasing, no roundups, no immobilizing of the animals. The quarry barely noticed that they'd been struck—perhaps feeling the equivalent of a bite from a stinging fly.

Colorado has been showing the way on contraception, not just in the Sand Wash area but also in Little Book Cliffs Wild Horse Range, northeast of Grand Junction, where there are 150 or so horses, and in the Spring Creek Basin, southwest of Montrose, which has 50 to 60. "In Spring Creek, we have a volunteer down there, who is darting 90-something percent of the mares," Fox said with pride. "We've got that down. The volunteer there is just damn good." In Little Book Cliffs, Fox noted that Friends of the Mustang in Grand Junction "have a very successful darting rate." Fox said that the BLM and the volunteers did supplement the darting program in the Little Book Cliffs region in 2013 by removing more than a dozen horses from the range by baiting them in a corral his team set up. "We did it all in-house with a volunteer group helping us, and when we took fourteen horses to town, we vaccinated, wormed, freeze-marked and entered them into the system and adopted every one of them out and not a single one of those horses went into the long-term holding pipeline."

After we darted the two mares, we hopped back in the vehicles in search of more horses, as daylight faded and the high temperatures receded to a far more comfortable range in the mid-80s. It took us another twenty minutes to find more horses—again about two dozen in a band. Just as before, we pulled off the dirt road, and Trueblood got out her binoculars and spotted just two more mares needing treatments. We zigzagged out to the herd, and Fox had one horse lined up and took his shot. Nothing. He'd forgotten to insert the dart and had fired a blank. He laughed out loud right along with us. Reclaiming his dignity, the former military helicopter pilot reloaded and had another dart in a mare's rump within five minutes. Trueblood was true on her second shot of the day, and there were now a total of four more horses that wouldn't see a foal ten months ahead.

Joining this humane hunt fortified my view that fertility control is a viable alternative to the unending and bankrupting cycle of rounding up and removing horses from public lands. For that reason alone, it makes great sense to grab the dart guns and maintain the ledgers of treated and untreated mares. But standing out in the vast reaches of the Sand Wash Basin reminded me that we've got so much backwards in our management of free-roaming animals on our public lands. While the BLM is still putting the bulk of its money into roundups and removals of wild horses, ostensibly to protect the range, the agency is welcoming millions of exotic animals onto those same lands every year. For every wild horse on public lands in the West, there are sixty-seven cattle and sheep. In fact, more than four million of these domesticated, nonnative species roam mainly BLM and US Forest Service lands, with ranchers paying below-market rates for the privilege. Only in the upside-down world of western politics are horses viewed as overabundant and in need of more severe control.

What's needed, in truth, are roundups of the cattle and sheep on public lands, and cease-and-desist orders to the ranchers once the animals are removed. The few ranchers who benefit from using our public lands deduct far more than they produce, diminishing the experience for all who use those lands for hiking, climbing, and wildlife watching, and creating havoc for wild horses and other wildlife. Cattle and sheep are drawn to riparian areas, and they gulp down fresh water, denude grasses, and compact the soil in areas that are vital to all wildlife. There is an enormous body of research about the effects of overgrazing and desertification caused by cattle and sheep on western lands.

On top of that, our own federal government, through the USDA's Wildlife Services program, is killing tens of thousands of coyotes, mountain lions, bears, and wolves on these same public lands—mainly in a savage war on native species waged to enable grazing by the ranchers' exotic species. Wildlife Services has been amassing a body count of 100,000 coyotes a year for decades on end by poisoning, trapping, and even aerial gunning them. Those females who make it out alive increase their litter sizes, and their offspring have higher survivorship rates—landing you essentially in the same place and requiring this annual treadmill of killing. That's as self-defeating as the roundup and removal program for horses: swallowing up financial resources, creating untold suffering, and requiring endless repetition just to stand in place. Taxpayers spend more than $60 million a year on the Wildlife Services program, with so much of that expense concentrated in the West.

And we're doing all of this to secure just 2 percent of the beef we eat in the United States. As for mutton, it is exceedingly rare on our food menus today, and the public lands sheep ranching program is designed more to provide a subsidy for the producer than a necessity for the consumer. It's crony capitalism: the government catering to a small, powerful group of people who represent no one but themselves, and the whole of society bearing the costs unapologetically shifted on to them. As with so many problems in the way that we treat animals, what's needed is proper accounting, so that we can make more rational policy choices once all the costs and consequences are tabulated.

In a broader sense, it's time to reevaluate our response to resource conflicts involving wildlife and people. As a nation, we've never been without these conflicts, and they're bound to increase with a human population of 320 million and substantial protected areas that allow wildlife to thrive. Mass removal and slaughter of animals was once a socially acceptable means of scouring away the problems. But that's no longer the case. We know too much about the characteristics, families, and social lives of animals to kill them without a trace of concern or conscience. Dealing with human–wildlife conflicts requires serious moral consideration, and fortunately, our creativity and ingenuity now gives us more options. Solving conflicts in a humane way is the right thing to do; it's more practical than ever; and it's increasingly supported by a wider range of stakeholders.

Rather than rounding up and gassing Canada geese in the East, where the birds flock to manicured golf courses or business parks, we can now addle recently laid eggs and keep their populations in check. In other parts of the country, municipal leaders and businesses fret about pigeons that, like Canada geese, leave droppings that people find aesthetically unpleasing and even unhealthy. Erick Wolf, the founder of Innolytics, sells a product called Ovocontrol, which is a fertility control substance tailored to the birds. It's a form of bait treated with a substance that prevents reproduction, and as with addling, it's an alternative to gassing or poisoning, which is a particularly awful way for these creatures to die. Wolf has won contracts in Chicago, Minneapolis, and other municipal centers to humanely resolve pigeon conflicts.

For state wildlife agencies dealing with beavers damming areas and creating flooding, the old way was to use a body-crushing trap, known as a Conibear. But today, Skip Lisle with Beaver Deceivers International and other experts in beaver management have developed bafflers, which allow water to flow through the dams yet don't trigger the beavers to rebuild and stop the water flow. That creates an enduring and cost-effective solution that eliminates the need for lethal trapping year after year when other beavers move into the habitat and reengineer the problem.

Increasingly, for every kind of conflict or nuisance problem with an animal, there is a solution available or in the making. At that point, it becomes a matter of making obvious moral and economically favorable choices, and not defaulting to the inhumane methods out of historical inertia and obedience to routine. Just as we stopped rounding up dogs and cats and gassing them in shelters, we'll stop rounding up wild animals and poisoning, trapping, or shooting them to deal with problems that are mainly of our own making. For every human–wildlife conflict, too, there's a business opportunity for resourceful entrepreneurs. Imagine the economic potential for businesses able to humanely control rats in cities, feral pigs in rural areas, and birds like starlings everywhere—in each case preventing problems before they're out of control and mass killings seem like the only solution.

In the Sand Wash Basin, as the sun slowly sets behind the mountains that frame this vast open range, you cannot help but feel something magical about seeing wild horses on this extraordinary western landscape. Of all of the species that helped the settlement and expansion of the United States, no species played a bigger role than the horse. These animals altered our notions of time and space, and it was on their backs that we explored our frontiers, settled the West, moved our armies, and carried our goods. Of all the creatures to whom we owe a debt of gratitude in this nation, the horse stands first in line. It seems a small repayment to them—the millions we keep in our pastures and use for recreation, and the wild horses that so many of us admire living on public lands—to treat these creatures with respect, decency, and mercy. For the ones still wild and free, it was seeing them where they belong, in all their elegance and exquisiteness, that motivated people to pass a law to protect wild horses nearly a half century ago. That same image of horses should inspire us today to do our part, leaving them in relative peace on patches of American land, ready to inspire the next generation with their unmatched majesty and untamed beauty.

CHAPTER SEVEN

Global Growth Stocks: Elephants, Lions, Great Apes, Whales, Sharks, and Other Living Capital

"Can he who has discovered only some of the values of whalebone and whale oil be said to have discovered the true use of the whale? Can he who slays the elephant for his ivory be said to have 'seen the elephant'? These are petty and accidental uses; just as if a stronger race were to kill us in order to make buttons and flageolets of our bones; for everything may serve a lower as well as a higher use. Every creature is better alive than dead, men and moose and pine-trees, and he who understands it aright will rather preserve its life than destroy it."

—HENRY DAVID THOREAU, *The Maine Woods*

Saving the Giants of the African Economy

When I saw Simon Belcher in 2014, he was just regaining the feeling in his arm, but he had already been back in the bush leading wildlife safaris. Belcher and his wife Amanda operate African Royal Safaris, taking high-end customers on trips into some of the wildest areas in all of Africa. But Belcher didn't sustain an injury from an angry Cape buffalo or a lion; he sustained his injury on his first-ever visit to the Westgate Mall in Nairobi, in September 2013.

Simon and Amanda, both Kenyan citizens, planned to see a movie and get a bite at a Japanese restaurant. Just after walking into the shopping area from the top level of the garage, they heard what sounded like firecrackers. "I thought it might have been an Indian wedding," Simon explained. Then people started running.

Simon and Amanda fled back to where they had come from—the top level of the garage where they'd parked. Two Somali men with AK-47s were walking up the ramp, shooting people and positioning themselves to prevent anyone from escaping, since there was only one way out of the garage. Hiding until help arrived was the only option. Simon took cover underneath a Land Rover, along with a boy and his nanny, and Amanda placed herself next to two men under a Toyota.

"I turned my cell phone off, but others near me had theirs on, and that's what drew the al-Shabaab attackers to us," Simon told me. "One of the men saw me, and then himself took cover behind a car. Only later did I realize he had thrown a grenade just five feet from me, but it didn't explode. He realized it hadn't gone off, looked at me again, and that's when he shot me. I was hit in the arm and the chest. I played dead so he wouldn't shoot me again."

Hours passed without any intervention by Nairobi police or the Kenyan military. The first help came in the form of a private group of Indian sharpshooters who got into the garage area and directed the victims to a safer area that they could defend. Simon said that a former Irish Army Ranger also guided people to safety—he and the Indian sharpshooters putting their lives at risk to help others. By this time, the attackers had retreated to an area inside the mall, and the rescuers helped guide Simon and the others to safety, and then to an ambulance. Simon told me that when he got to the hospital, he was euphoric, knowing that both he and his wife had survived the onslaught. Doctors realized, however, that he was not okay. They immediately inserted a tube through his ribs to drain fluid from his lungs. Just a few more minutes without treatment, and he would have been dead.

In the end, attackers killed more than 60 innocent civilians in the assault on the mall and wounded 175 others. When we met, Simon told me that Kenya is facing an existential threat from al-Shabaab and other Somali-based terrorist groups. The terrorists are killing civilians in brazen attacks, trying to create a climate of fear. They're also involved in slaughtering elephants and trafficking the elephants' ivory to finance their murderous operations. In the process of going after people and wildlife, they are threatening the tourism industry that is crucial to the Kenyan economy.

Kenyans realize, Simon told me, that the well-being of wildlife is essential to the future of the nation. Second only to agriculture, wildlife tourism is one of Kenya's largest industries. "The Maasai don't hunt [elephants], but they didn't care too much about Somalis coming to kill elephants," Simon explained. "But when they realized that the elephants translate into tourism and dollars for their communities, then they worked to protect the elephants."

There is a ready explanation for the take-your-breath-away appeal of Africa's wildlife and the existence of a business like Royal African Safaris. Elephants, rhinos, hippos, and giraffes are the largest of the world's land mammals, with each species displaying otherworldly or even prehistoric-looking characteristics. Then there are the millions of wildebeest, zebras, buffaloes, gazelles, and other hooved animals marching literally to greener pastures, as if in a parade, with an urgent purpose during the summer migration. Tracking their every movement with piercing eyes is the greatest variety of large predators in the world—lions, leopards, cheetahs, hyenas, painted dogs, crocodiles, and dozens more. In prides, packs, or as solo artists and by long chases, short bursts, or amphibious attacks, they chase and kill quarry often many times their size. It's the closest thing the world has to a real-life Jurassic Park.

I got a firsthand view of this drama a decade ago on my very first wildlife safari in East Africa. As our group was traveling through the Serengeti, we fixed our gaze on a herd of three hundred or so buffalo, about one hundred yards away. Their big, black, muscle-bound bodies rippled in the sunlight, their heads down as they crushed shoots of grass under their hooves while ripping at other blades with their teeth and lips. They'd look up occasionally, massively confident and without wariness, grinding tufts of grass with tails swiping and ears flicking to throw off swarms of insects. We'd stopped to watch them from a plateau, with a downward-facing view of the valley they were in, and had only been there five minutes before we noticed a pride of half a dozen lions, each crouching close to the ground and taking up positions on two sides of the herd. Once in place, they settled in and waited for an opportunity to bring down a buffalo. We turned off the vehicles and settled in too, expecting violence.

It didn't take long before we saw an adult buffalo and calf who had strayed from the herd by at least thirty yards. It didn't escape the lions' notice either. Every one of us, we later told ourselves, had been muttering to ourselves for the mother to rejoin the herd—not because we were picking winners and losers, but simply from an instinctive feeling of empathy. It didn't take much imagination to know what was about to unfold.

Lions will starve and die if they don't drag down buffaloes, zebras, and other creatures at a consistent clip, and there is no malice or cruelty in their predatory ways—just the necessities of life, absent any ritual, remorse, or celebration. Yet who among us doesn't feel pity for prey animals and their families and herd mates? It's true that these routine bursts of violence result in the taking of life, but they help sustain it too. And the outcome of such encounters is seldom preordained, with prey often mounting a successful defense and escaping.

In the scene my friends and I were watching from the plateau, the lions seemed to be sending signals to each other, like players and coaches in a baseball game, though with fewer obvious cues. After a few tense minutes, four of the lions sprang from the tall grasses and ran at the mother and calf. A pair of lionesses jumped on the calf, who was slightly bigger than the lions. Two others charged the mother, who met them halfway as she rushed toward her calf. They pounced on her, but she shook them off and charged the other attacking lions. The remaining two lions made a bluff charge at the herd—apparently as a way to send the rest of the buffalo in the opposite direction and further isolate the quarry and give their pride mates time to do their cutting, with claws and teeth, at hidebound flesh. The two lions leading the attack released the calf when the mother came at them, but there were too many for her to fend off, and others jumped on the younger animal.

Meanwhile, the herd of buffalo was reforming its lines. But the animals hesitated in charging the lions and rescuing their herd mates. We were all beside ourselves, wondering why the herd didn't take advantage of its superior numbers. The lions, for their part, showed no hesitation. They now had the calf down, and they were again also on the mother, who was working to shed them so that she could make a furious run at her calf's attackers and use her powerful neck and horns to rescue her baby. The lions dragged the calf down into a gully, and we lost sight of them. By that time, however, the herd had gathered its collective resolve and finally made a lumbering yet determined charge, at least one hundred strong. The lions scattered, overwhelmed by the sheer mass of muscle and horns coming at them. The lead buffalo parted to create space, cossetted the mother and perhaps the calf in the middle of the herd, and then closed ranks around them. We weren't able to determine the calf's fate but had the feeling that she'd been badly injured yet perhaps survived. What started as sympathy for the plight of the mother and calf turned into concern for the difficulties that the lions face—with one buffalo making a bad decision, and with the pride seizing that opportunity, but still unable to finish the job. It was a lot of effort and risk with no reward.

The whole clash and struggle couldn't have been more dramatic or vivid. If I needed any explanation for why millions of people trek to Africa's wild places, here it was. More than ever, the political and business leaders of African nations are recognizing the remarkable wildlife assets within their borders, and they're taking steps to safeguard them as a way of both doing the right thing and boosting their economies. Kenya alone has fifty-one terrestrial national parks and reserves, worth $48 million a year in gate fees and billions more in related commerce: travel, lodging, guiding, food, photography, and other spending. Wildlife tourism yields 70 percent of tourism earnings, 25 percent of Kenya's GDP, and more than 300,000 jobs.

Kenya has forbidden sport hunting since 1977. The county's leaders realized that, while hunting concessions could generate some money, that activity would remove many of the biggest and most charismatic wild animals and alter the behavior of survivors—making them more skittish and less tolerant of people. If the animals bolt at the sight, smell, or distant sound of a human—and who can blame them when you don't know if it's a friend or foe approaching—there's much less in the way of good wildlife watching to be had. Several other countries, including Zimbabwe and South Africa, think they can have it both ways—trying to lure wildlife watchers even as they auction off wildlife to trophy hunters.

In July 2015, Walter Palmer, an American dentist working with two professional guides in Zimbabwe, worked to lure Hwange National Park's biggest black-maned lion outside of its protected confines and onto private land by tying a dead elephant carcass to a truck and dragging it along the border area to create an irresistible draw. At about 10 p.m., the lion strayed from the park to gorge on the elephant, and one of the hunter's guides threw a spotlight on the animal. Palmer let loose with a broadhead arrow that struck and injured the lion, named Cecil by the Oxford University researchers who had been studying him and the other lions of Hwange. Palmer and his two guides chose not to give chase, presumably for fear of tracking an injured lion in the bush during darkness, so they went back to their camp and slept through the night. The lion, undoubtedly injured and weakened, presumably got no sleep and suffered for at least a dozen hours before Palmer resumed the hunt the next day and finished him with a second arrow. It was then they realized Cecil was wearing a radio collar, a marker that the animal was part of the Oxford lion study. Instead of reporting the incident, the threesome reportedly tried to cover up the killing by moving the collar away from the kill site and then destroying the tracking device. (Palmer maintains he did nothing illegal, but this wasn't his first alleged attempt to cover up a wildlife crime. He had a felony guilty plea on his record, after illegally slaying an enormous black bear in Wisconsin years before, then attempting to bribe his guides to lie about the location of the kill.)

Once news of the killing of this well-known lion broke, there was a global furor. Palmer hadn't just needlessly slain a lion. He had robbed a park of one of its central characters, diminishing the experience for visitors who traveled there precisely to catch a glimpse or take a photograph of Cecil. "This trophy hunting is destroying our wildlife, and for what really?" Zimbabwe Conservation Task Force chairman Jonny Rodriguez told a newspaper in the country. A 2013 study revealed that trophy hunting in 2011 generated just 0.2 percent of the nation's GDP, while nature-based tourism generated 6.4 percent. "We have lost a lion which marketed our country and provided many with treasured memories," added Rodriguez. "Cecil was killed for only US$50,000 [an amount that Palmer disputes], and yet he was worth more than a million dollars." That's a conservative estimate, since Cecil was perhaps the top draw at Hwange. Just one night at a five-star lodge near the park was priced at $763 per person, for the sleeping quarters, food, and wildlife-watching excursions—an enormous investment of foreign cash in a nation racked by poverty.

Americans like Walter Palmer kill more than 700 African lions and about 600 elephants every year, doing so in fewer than a dozen countries, with Zimbabwe one of the top three destinations for American trophy hunters. Palmer's guide said the Minnesota-based hunter also wanted to kill a trophy elephant on that trip, but they couldn't find one large enough to his liking. Foreign tourists like Palmer not only hand over hefty fees to the guides, but also to governments and wildlife officials, in a far-flung pay-to-slay scheme. There is a long list of cases of these international trophy hunters bending or breaking the rules—using unfair means or killing in protected areas, for example—to kill prized animals in order to secure a place in the record books of Safari Club International or, for bow hunters, the Pope and Young Club. The trophy hunters are drawn precisely to the nations that allow something of a free-for-all in the killing of rare species. The United States suspended any imports of sport-hunted elephant tusks from Zimbabwe or Tanzania starting in 2014 because of the nations' widespread corruption, lack of effective enforcement, and mismanagement of wildlife. It was only after the global furor over the killing of Cecil that Zimbabwean officials feigned indignation and took some remedial actions, presumably to mitigate some of the public relations damage from the incident. Zimbabwe restricted for a few weeks the killing of elephants, rhinos, and lions around the national parks where Cecil lived, but then lifted the ban as the global spotlight dimmed. The government also arrested the local guide and landowner and called on the United States to extradite Palmer. But then weeks later, government officials said they wouldn't charge Walter Palmer and indicated they'd welcome him back as a tourist. Three months later, a German trophy hunter came to Zimbabwe and paid $60,000 to kill the biggest elephant seen on the continent in decades. If there was any hope for serious reform in the southern African nation, Zimbabwe's ninety-one-year-old dictator, Robert Mugabe, had long ago put that notion to rest: not long before the killing of Cecil, the tyrant threw himself an extravagant birthday party in an economically disadvantaged nation, feasting on an elephant killed just for the occasion. In order to protect wildlife and to drive humane and sustainable wildlife-based tourism, we not only need to root out the terrorists, but also retire the despots. The humane economy is resilient and adaptive, but it depends on the rule of law and the adherence of leaders to the tenets of a civil society and democratic conduct. Dictators who eat elephants aren't part of such an enterprise and never will be.

While the US government has publicly condemned Zimbabwe because of its disgraceful leadership, trophy hunters see opportunity in political chaos. In places like that, the business of trophy hunting is transactional, with killing opportunities all but guaranteed, along with the accolades to follow in the world of competitive global hunting—mainly from Safari Club International, with its twenty-nine "hunting achievement" and dozens of "Inner Circle" awards. To secure all of these awards, a trophy hunter must kill as many as 322 different species and subspecies, and the killing is too often done in countries where the rule of law is nonexistent and elections are a sham. Among the most coveted of the awards is the "Africa Big Five," which Safari Club members claim by killing an elephant, lion, rhino, leopard, and Cape buffalo. Kill all five and your name is enshrined in the record books—and what much of the world thinks of as a killing spree is, at least in the small fraternity of the Safari Club, treated as a lifetime accomplishment.

The list of Safari Club record holders would give nations wanting to protect their wildlife a head start in developing a "no entry" list. Even if they come with check in hand, and even if they are killing lawfully as required by Safari Club rules, it's prudent to keep these people out of the country to prevent them from looting its wildlife. In America, that same list could give US Customs officials and Fish and Wildlife Service enforcement officers, sorting through a vast number of big boxes and carry-ons with wildlife parts, a few good leads to apprehend wildlife traffickers at our ports of entry. And if that list ever gets out in a public way, it could be a great asset to consumers and voters, should the names of politicians, CEOs, dentists, or others become widely known. Resigning from office for the elected officials or, as Walter Palmer did, even temporarily closing up his business practice and going into hiding might be the right way to handle the problem. That would allow them plenty of time for self-examination to find out why they spend so lavishly, travel around the world, and get such a psychic high from killing rare animals and putting their remains on permanent display in their trophy rooms or home museums.

Thanks to Palmer, for at least a little while the trophy hunting subculture came to light, and the public didn't like what it saw. It prompted some in the corporate world to better align their policies with public sentiment. Since the enterprise of globetrotting trophy hunting depends on commercial transportation, the HSUS and other animal-protection groups called for the airlines and airfreight carriers to stop enabling the enterprise by shipping the heads and other parts from any of the Africa "Big Five." Lufthansa, KLM, Singapore Airlines, Emirates, and Virgin Atlantic already had policies in place, and within days of Cecil's killing, British Airways, Air France, Brussels Airlines, Iberia Airways, IAG Cargo, Qatar Airways, Qantas, Virgin America, Virgin Australia, and Etihad announced or confirmed similar policies. The HSUS then stepped up the call in the United States, and within days, Delta announced it "will officially ban shipment of all lion, leopard, elephant, rhinoceros and buffalo trophies worldwide as freight." Just as one airline follows another on pricing, more carriers imposed similar restrictions. Within twenty-four hours of Delta's announcement, American, United, and Air Canada followed suit. DHL was next and the campaign continued with the few remaining outliers. "Trophy hunting feels like a relic of a bygone era when people were conquerors, rather than stewards, of their environment," wrote Richard Branson, founder of Virgin Atlantic, in an August 2015 blog post titled "Big Game is worth more alive than dead." Branson then called upon all airlines and freight carriers worldwide to develop "a strict ethical cargo policy to clearly identify shipments not acceptable for carriage." The humane economy has its thought leaders, and the head of Virgin Atlantic is certainly one of them.

We argued to airline executives, especially those not as versed as Branson on the subject, that they should not provide a getaway vehicle for the heist of Africa's wildlife by these killers. With thousands of people writing to them, they heard the roar of the public. By the end of August, a little more than a month after Palmer's crime came to light, more than forty airlines and freight carriers got on board. And the decision makers didn't require the CFO to dig into the numbers and see if this was a financially prudent move. It's plain to see that the number of trophy hunters is shrinking, and the number of wildlife watchers swelling. For every trophy hunter who might fly to Africa in search of gunning down a living creature, there are thousands of wildlife watchers with cameras in search of a full-body picture of a glorious creature. With the trophies out of view in the cargo hold, that was fine when the public was unaware. But with the klieg lights on, post–Walter Palmer, suddenly there was no escaping the underlying economics for the airlines. The legroom and the departure times mattered, but so did the steerage now.

Then the trophy hunters sustained some big blows with the closing of markets to their trophies. Responding to pleas from the legendary actress Brigitte Bardot, France banned any imports of lion trophies. Just weeks later, in December 2015, the US Fish and Wildlife Service, acting on a petition from HSUS, declared African lions "threatened" or "endangered" across their range. This designation—driven by the decline in the population of the great cats due not only to trophy hunting but also to habitat loss, poaching, and other threats— means that trophies can be imported into the US only if the activity enhances the survival of the species. The federal government specifically said that shooting lions in captive hunts could not possibly meet that standard— and American trophy hunters took more than half of the 720 lions they killed and imported from fenced enclosures in South Africa where the animals were bred and then shot in guaranteed kills. With the captive hunts dominating the lion-hunting business, and with corrupt management of lion-hunting programs in Tanzania and Zimbabwe accounting for much of the rest of the enterprise, the new federal rules would likely slow imports of lion heads and hides from a torrent to a trickle. The airlines were making it difficult to transport the trophies, and now France and the United States—and Australia earlier in the year—established policies that severely restrict or forbid the imports by air, land, or sea. The few remaining countries allowing lion trophy imports would be under pressure to adopt their own strictures, threatening to put an end to the entire despicable and bloody business of guided and captive hunts for the king of beasts.

According to a 2013 report by the World Tourism Association, wildlife-based ecotourism delivered an estimated $34.2 billion in tourist receipts, while a different study of nine countries that offer trophy hunting found that, in 2011, tourism contributed 2.4 percent of GDP, and trophy hunting only 0.09 percent of GDP. In South Africa, the biggest trophy hunting country on the continent by a long shot, wildlife tourism brings eighty-five times the revenue that trophy hunting does. Airline executives realized that if they persisted in shipping that kind of freight, they'd lose business and tar their reputations. About 80 percent of tour operator trips to Africa are for wildlife watching. Almost overnight, thanks to the air carriers doing the right and responsible thing, trophy hunting got even more expensive and complicated for its human participants. It wouldn't be enough any longer simply to pay off corrupt wildlife officials or unscrupulous hunting guides in struggling nations—they'd now have to figure out a way to get their trophies back home. In the wake of the killing of Cecil, and the US rule forbidding elephant imports from the country, trophy hunting receipts in Zimbabwe declined by 30 percent in 2015.

There's mounting pressure for African nations to align their policies with non-threatening wildlife tourism. Because of trophy hunting bans in certain countries, or import restrictions placed by the US government, American hunters can now only import African elephant trophies from Namibia and South Africa. But even those two outliers could be soon closed off to trophy hunting for elephants too. After the Cecil killing, a group of US lawmakers demanded a ban on any imports of threatened or endangered species, or any species proposed for protection under the Endangered Species Act, with a bill called the Conserving Ecosystems by Ceasing the Importation of Large (CECIL) Animal Trophies Act.

Forward-thinking African nations have already charted a new course and put aside wildlife slaughter as an economic activity. In 2013, recognizing the vastly larger market of wildlife-friendly tourists compared to wildlife-killing tourists, the president of Botswana announced a ban on virtually all trophy hunting—even though the country had been the biggest elephant hunting country on the continent, previously accounting for nearly half of all trophies coming into the United States. In making the announcement, President Seretse Khama Ian Khama noted that "if we do not take care of our animals, we will have a huge problem in terms of tourism." Tshekedi Khama, Botswana's environment minister, said that nonlethal tourism was better: "Hunters only employ people during the hunting season. [Tourism] is throughout the year—that's why we prefer it." In 2015, Botswana began advertising in *National Geographic* magazine, showing off two beautiful lion cubs and a double-barreled message:

> "Experience true ecotourism in a visionary country that outlaws hunting and believes that only a camera should ever be aimed at a lion. Discover our award-winning ecolodges on a wildlife safari dedicated to protecting nature."

It's hard to imagine that a compassionate society, concerned about vanishing wildlife, would continue to tolerate the far-flung missions of these self-indulgent international trophy shooters. At a time when Western nations are asking poor Africans to stop killing elephants for their ivory, is it too much for rich Americans to stop killing elephants for their tusks and heads? Some of the poachers, at least, can plead poverty as their excuse—though even in Africa, that doesn't go far and poachers are despised for their cruelty, theft, and disregard of the common interests. But American trophy hunters do their killing purely out of privilege and greed, as well as for whatever pleasure is gained from terrifying and killing elephants and other creatures. When we ask African leaders and citizens to show restraint and stop killing and to invest in ecotourism, but don't place legal restraints on the worst, most gratuitous offenders among our own ranks, whatever moral authority we might have takes a hit. And what kind of hardship is it anyway to ask a fellow from Dallas or Minneapolis not to travel eight thousand miles for the sole purpose of shooting animals in what amounts to a headhunting exercise? Why not spend your money on doing some genuine good, rather than invading a beautiful, faraway place to mete out death and destruction? And the picture taking, with the hunter sitting or standing atop the slain animal, smacks of a bizarre sort of voyeurism and an earlier era of colonialism.

Gabon, which has a large share of all surviving forest elephants on the African continent, has, with Kenya, long forbidden trophy hunting. In 2002, then-President Omar Bongo created a new national park system, including thirteen units that cover more than 15 percent of the country—key areas for the sixty thousand or so forest elephants in Gabon. In a country of more than a million people, current President Ali Bongo Ondimba has set up highly trained anti-poaching units to stop the large-scale killing of forest elephants. In 2014, he designated a new network of ten marine parks, covering more than eighteen thousand square miles, to safeguard whales, sea turtles, and other marine species that inhabit the country's coastal and offshore ecosystems. Gabon is one of the only places on earth where elephants and whales can see each other in the wild, the world's largest mammals swimming close to shore and its largest terrestrial mammals occasionally leaving the forests to spend time on the beach. The presidents of Gabon and Kenya know they've got something special—and they've set a precedent by seeking to protect it forever.

Along with the national parks, marine reserves, and other areas protected by government decree, there are dozens of private reserves throughout Africa where people can take in amazing wildlife-watching experiences. I visited one such preserve in Kenya: the Lewa Wildlife Conservancy, a 55,000-acre privately owned sanctuary teeming with wildlife, including a noticeable number of highly endangered black rhinos. During my trip there, Ian Craig, whose forebears set aside a portion of this land in 1924 as a protected area for black rhinos, brought a group of us right into the middle of a herd of a dozen or so elephants who'd fanned out over a patch of land to feed at dusk. With the sun setting and the midday heat lifted, they were using their big bodies to knock down trees and deploying their trunks to rip off branches so they could swallow up the fresh greenery. Their display of strength and appetite filled the air with the scent of fresh pulp and bark, and we spoke in hushed tones so as not to miss the sounds of snapping branches and crashing trees. You can understand, after witnessing this display, how elephants are able over time to alter their environment, as beavers do in North American ecosystems by felling trees and creating wetlands. As we intruded on their dinner, one of the big females put herself between us and the rest of the herd, ears wide and eyes fixed on us. I was not the only one among us to wonder if she knew she could topple our jeep even more easily than the trees her smaller family members had just flattened. She let us be, reminding us that despite so many bitter experiences with humans, even the most powerful animals seldom wish to do us any harm.

Within their ecological communities, elephants are a keystone species who leave their mark on the forests and grasslands. In habitats big enough to sustain elephants, you'll likely find the full complement of other African wildlife—with the exception perhaps of rhinos, who, because their horns are coveted in pockets of Asia, have been wiped out across much of their range.

Elephants are a keystone species in another way too—they are at the top of the economic pyramid for the wildlife tourism industry, generating billions of dollars in commerce. "A single dead elephant's tusks are estimated to have a raw value of $21,000," according to a report by the David Sheldrick Wildlife Trust. "By comparison, the estimated tourism value of a single living elephant is $1,607,624.83 over its lifetime to travel companies, airlines, and local economies thanks to tourists willing to pay generously for a chance to see and photograph the world's largest land mammal. That makes a living elephant, in financial terms, as valuable as seventy-six dead elephants." One way to read these numbers is as a measure of consumer power: by not buying ivory goods and by traveling only to African countries where wildlife is protected, each of us plays an important role in advancing the humane economy.

The team at the Lewa Wildlife Conservancy works with communities adjacent to the reserve, delivering education curricula, water filtration systems, and jobs as wardens, guides, and sellers of artisanal products. The well-being and livelihood of so many people depends, in part, on the continued flow of tourists coming to see the preserve, enabling members of the community to view wildlife as an asset rather than a liability. And while the wildlife protection policies of a nation matter immensely, it is essential that conservationists also build support among the communities living adjacent to the wildlife. The communities can be, based on their proximity, wildlife's best insurance protection or its worst nightmare.

As with the contradictory attitudes we encounter toward animals in the West, countless Africans still kill wildlife and consume the meat or sell it. People kill an estimated half a billion animals a year or more for bush meat—a staggering toll that reminds us precisely why so many species of wildlife are in danger of extinction. The bush meat crisis is most acute in Central and West Africa, but it's present wherever people with access to wildlife struggle for survival. The problem has been made worse by investments in extractive industries in Africa, putting loggers, miners, and other resource seekers in places where animals had long been isolated from human intrusions, principally in largely inaccessible forests. Now there are roads, camps, and people—and they've often proved a deadly combination for wildlife.

Occurring on this vast scale, the bush meat crisis is a threat to the future viability of wildlife watching enterprises in Africa, since elephants, chimps, giraffes, and every species you can think of is killed and consumed. But it's also a public health threat, since 60 percent of diseases that afflict humans start in animals and jump the species barrier. It's thought that Ebola, the deadly zoonotic disease that killed more than eleven thousand in Liberia and other West African nations in 2014, originated in people who were killing and eating wildlife. As a means of promoting prevention, local rap artists there sang "don't touch your friends . . . don't eat something dangerous." A generation earlier, AIDS likely arose as a consequence of people killing and eating wild primates infected with a deadly variant of the virus. The effect of AIDS on Africa is still being felt profoundly decades later, with thirty-five million people on the continent infected, including a third of women in some southern African countries.

In fighting the bush meat trade, Ian Saunders of the conservation group Tsavo Trust says it's critical to win "the hearts and minds" of local communities. Many of the wildlife safari businesses operate by this credo, turning these communities into activists against the bush meat trade. During my time in Kenya, with the support of Anne Kent Taylor and the luxury tourism company Abercrombie & Kent that her family operates, we conducted patrols, walking through "protected areas" with community members to spot and uproot snares, which snap shut on the foot or neck of any creature unlucky enough to trigger the device. One young man on our de-snaring patrols, Josphat Ngonyo, had formed a nationwide animal welfare group in Kenya with anti-snaring as a signature issue. His African Network for Animal Welfare has grown since then, and it advocates for public policies to protect wildlife and domesticated animals. Josphat reminded me that if the wildlife are gone, tourists will stop coming, and the people of these communities will suffer most.

Dame Daphne Sheldrick leads the group named for her late husband, the David Sheldrick Wildlife Trust, which takes in orphaned elephants, typically after their adult family members have been shot, and it too is a beacon for wildlife protection in Kenya. At the facility, caretakers hand-feed the baby elephants bottled milk and even sleep next to them. At first, Sheldrick says, "they think we're the enemy." But before long the elephants are putty in the hands of their gentle caretakers; the facility is perhaps unique in the world in offering this kind of one-on-one, round-the-clock personalized care for these juvenile giants, showing them the kind of tenderness and attentiveness that only a mother could surpass. When you see the commitment of these workers, how could you possibly tolerate people who'd shoot the elephants or saw off their faces to obtain their tusks? When you watch tenderness up close, you get a sense of the deep reservoir of compassion for animals among so many Kenyans. And when Sheldrick's team rehabilitates an elephant and puts one of them back into the wild, repatriating the animal to his or her range and herd, she is, in practical terms, putting money back into the ecosystem, restoring life but also fortifying one of the most important industries in Kenya.

Some of Daphne's orphans come from Tsavo National Park, home to Kenya's largest populations of elephants. This protected area is a hub of wildlife tourism, with famed, long-studied lions and some of the largest elephants on the continent. I was fortunate enough to travel from one end of the park to the other and to see so many animals native to the arid parts of East Africa. So it was like a punch in the gut when I learned last year that a poacher, using a poison-tipped arrow, had killed Tsavo's most famous elephant—an enormous male named Satao, who sported sweeping, seven-foot-long tusks. All it took was one renegade poacher to take this creature's life and rob everyone else of the joy of being in his presence. In fact, on the very day that Cecil was killed—robbing Hwange of its most famous lion—poachers reportedly brought down five other Tsavo elephants in an act of ecological and economic sabotage.

Kenya had an estimated 167,000 elephants in 1979, but now has fewer than 28,000. Critics sometimes suggest that it's a choice between poor people making a living by poaching elephants and selling the ivory, versus protecting the animals for their own sake. But it is by keeping elephants alive that communities can deliver education programs for kids, bring women out of the home and into the workforce, and support the infrastructure to keep the tourism economy going. A slaughtered elephant, even in the coldest terms of economics, is not a resource used but a resource squandered.

Where there is poaching, there is generally a subculture of mayhem and anarchy. The terrorists and militants leaving a trail of human casualties in their wake have also looted entire nations of elephants, with devastating economic effects. When was the last time you ever heard of anyone going on a wildlife safari to Somalia or Sudan or Angola? Years after Angola's long and bitter civil war ended, the scarcity of wildlife has kept it off just about everybody's destination list. There were roughly 200,000 elephants in Angola decades ago, but today there are fewer than 2,000, in a country three times the size of California. The few surviving elephants are the ones with their sense of smell fourteen times more acute than a dog's, who have been able to sniff out the land mines planted during times of war. It's called the "Empty Forest Syndrome," and it's all too common where the ivory hunters, militias, and bush meat traders have left a trail of destruction—with some beautiful scenery intact but rarely a chirp from a bird, the call of a monkey, or a trumpet or a rumble from an elephant. When the wild animals are gone, or barely hanging on, why would tourists come and spend?

In the end, ecotourism is itself a competitive industry, and the countries doing things the right way, and keeping wildlife in the wild, are going to win the hearts—and dollars—of tourists. Never have we seen a case where the fortunes of nations, in terms of security, peace, and ecological health, have been so inextricably bound together.

More than ten million elephants are estimated to have roamed sub-Saharan Africa a century ago. Today, fewer than 400,000 remain, even as more people than ever want to see these majestic creatures moving freely in their natural habitats. It was a poaching bloodbath in the 1970s and 1980s—with perhaps 100,000 elephants killed a year during the worst of it—that led the global community in 1989 to impose a worldwide ban on the international ivory trade. Kenya Wildlife Service director Richard Leakey famously set a pile of confiscated tusks aflame as a symbol that he and his country would not be seduced into the short-term trap of trading in ivory, only to see global demand for the product eat up his nation's remaining elephants. Leakey wanted to keep tusks in the living tissues of elephants in the wild, because that's the wellspring of sustainable commercial enterprise involving wildlife.

The ivory ban of the late 1980s dramatically slowed the killing, but it was short lived, undermined by loopholes and one-time sell-offs of stockpiled tusks that only accelerated international demand. It was also compromised by conflicts that sent armed militias into the wild, giving rise to militarized poaching, as well as by the spread of Chinese business interests into Africa—enterprises that often informally serve as key brokers for ivory hunting and smuggling. While poaching today is a threat just about everywhere elephants live in Africa, a DNA analysis of seized ivory from shipments from throughout Africa shows it's principally occurring in two hotspots. The study found that "more than 85 percent of the forest elephant ivory seized between 2006 and

2014 was traced to the central African Tridom protected ecosystem that spans northeastern Gabon, northwestern Republic of Congo and southeastern Cameroon, and the adjacent reserve in southwestern Central African Republic." It also found that "more than 85 percent of the savanna elephant ivory seized between 2006 and 2014 was traced to East Africa, mainly from the Selous Game Reserve in southeastern Tanzania and the Niassa Reserve in adjacent northern Mozambique." Ranking 185th out of 187 countries on the UN Development Program's 2013 Human Development Index, Mozambique has been a terrible place to be an elephant in recent years. An estimated 9,345 elephants were poached out of a population of just 20,374 between 2009 and 2013 in the Niassa ecosystem. In Tanzania, which once had the largest elephant population in the world, at more than 300,000 just decades ago, government corruption and professionalized poaching has taken an awful toll. In the Selous Game Reserve and its surrounding ecosystem, the elephant population has fallen from 109,000 in 1976 to 13,084 in 2013.

These dire numbers should motivate governments in African range states, in the United States, and in the European Union to make far greater investments to fight the poachers. Mozambique and a number of other countries are taking action as never before. It should also motivate wildlife protection charities and philanthropists to fortify the ranks of park wardens, replacing their World War II–era weaponry with modern firearms and other equipment to give them decisive force against heavily armed poachers. Adding drones as a tool to spot the animals and the poachers offers the promise of zeroing in on the perpetrators. In July 2015, Dr. Paul O'Donoghue, a British scientist working with rhinos, announced the invention of the Real-time Anti-Poaching Intelligence Device (RAPID), which combines a GPS satellite collar with a heart rate monitor and video camera, allowing a signal to be sent immediately when an animal's heart rate and body chemistry signal there's a poaching incident in progress. At that point, it may be too late to save the animal, but RAPID may allow anti-poaching teams and national armies to pick up a hot trail on perpetrators rather than simply stumbling across a mangled carcass of an animal poached days or weeks before. The poachers have resorted to poisoning vultures, since they don't want the presence of these scavengers to signal to authorities that elephants and other creatures are dead, so these new technologies are especially important to law enforcement personnel.

Elephant poaching is deeply intertwined with lawlessness in Africa and the breakdown of civil society. The al-Shabaab, Joseph Kony's Lord's Resistance Army, and other terrorist groups often finance their murderous ways by liquidating wildlife and selling the parts to wildlife traffickers. "Ivory operates as [a] savings account for Kony," says the US State Department's Marty Regan. Inevitably, they are the first link in a long global supply chain that connects ivory to global markets thousands of miles away from the lands where the animals live. Some Chinese consumers, it's said, are so detached from the killing that they don't think you need to kill elephants for their ivory—believing that ivory comes from teeth that naturally fall out without harming the elephants. As I write this in 2015, a carved tusk in China goes for $1,300, and a pair of sizable tusks brings $50,000 on the streets.

The Chinese presence on the continent has been substantial and growing, with billions of dollars invested, hundreds of companies operating there, and perhaps more than a million Chinese nationals in Africa. In a 2012 cable posted on Wikileaks, the US ambassador to Kenya noted "a marked increase in poaching wherever Chinese labor camps [are] located." These people provide a ready means of getting goods back to China, including illegal ivory. "Poaching has risen sharply in areas where the Chinese are building roads," Julius Kipng'etich, then-director of the Kenya Wildlife Service, told the *Telegraph* in 2011. "Is that a coincidence? Ninety percent of the ivory confiscated at Nairobi airport is in Chinese luggage. Some Chinese say we are being racist, but our sniffer dogs are not racist."

But it's too simple to blame only the Chinese for the killing. There were 130,000 elephants in Sudan twenty-five years ago, and today there are just 5,000 or so between Sudan and the newly created South Sudan. Much of the wildlife there was simply consumed by the Sudanese armies and militias during the decades of civil war. Today, Sudanese bands, likely associated with the ruthless Janjaweed militia, are fanning out to the south and financing their terrorism by killing elephants. "Sudanese hunting expeditions are today operating more than 600 km outside North Sudan's borders into Chad, Cameroon, Central African Republic (CAR), and northern Democratic Republic of the Congo (DRC) in order to poach central Africa's remaining elephants," according to an exhaustive report from C4ADS and Born Free USA. Some decades ago there were perhaps a million elephants in the Central African Republic alone. Today, there are probably fewer than 100,000 elephants in the entire group of countries in central Africa—most of them in Gabon, where the president is arming wardens to fight poachers.

In the past decade, nearly a thousand park rangers have been killed worldwide while defending animals in the wild. Their courage is something to admire, and more nations are stepping up to safeguard their wildlife protection teams. In Chad, after Sudanese poachers ambushed and killed five rangers at dawn outside of their tents at Zakouma National Park in 2012, President Idriss Déby set ivory stockpiles aflame and committed military resources to aid national park personnel. In March 2015, Ethiopia's Deputy Prime Minister Demeke Mekonnen set ablaze his country's entire stockpile of six tons of ivory as " 'a vital stepping stone' in the implementation of more stringent laws against poaching." The following month, Malawi's President Arthur Peter Mutharika pledged to burn the nation's 6.6 ton ivory stockpile. "The destruction of the ivory stockpile is not merely an action of government to protect the future of the country's dwindling elephant population," Mutharika noted, "but also to send a strong signal to the rest of the world about Malawi's commitment to the fight against wildlife crime." The same month, President Denis Sassou Nguesso of the Republic of Congo presided over the burning of five tons of ivory there. When you have so many national leaders giving personal attention to this issue, you know it's not an abstract concern or a symbolic one, but a matter of fundamental importance to the economic future of their countries.

Yet the African leaders' efforts will be for naught if ivory-consuming nations don't work to dry up demand and strengthen international restrictions on the trade. Commercial interests in China and the United States have fed that demand by making and marketing trinkets and other common goods, or attaching pieces of ivory to musical instruments, guns, or knives. In the wake of the 1989 ban on international trade in ivory, domestic ivory markets withered without reliable supplies of ivory. Sales were largely restricted to pre-ban ivory and antiques, and the best evidence came in the form of ivory carving factories in China and Japan shutting down.

But ten years after the ban, China and Japan, along with southern African countries, collaborated on a plan to reverse the ban and revive the carving industry. Together, they lobbied for international approval for two sales of tons of ivory from four southern African countries. The influx of ivory in Asia revivified the carving industry, and when these businesses exhausted the ivory from the two sales, they set up channels to obtain more products through illegal means. They found ready trading partners in African militia groups who needed cash for ivory. The Chinese carving industry then started churning out massive quantities of statues and easily smuggled smaller items, ostensibly sold as "antique" ivory jewelry and trinkets, in order to feed the markets in China and the United States. The problem is that the legal trade provided a cover for the illegal trade, and as demand increased, so did the killing of elephants. The United States became the world's second largest market for ivory, after China.

But present-day consumers are connecting the sale of these products with the slaughter of the biggest land mammals in the world, and many are rallying to stop it by favoring easily obtainable substitutes. Now you cannot find a piano company anywhere that makes keys from ivory. EBay and other Internet sellers stopped any sale of ivory trinkets on their sites. Close the markets and the poachers will be left holding the bloodstained ivory in their hands, with no middleman to purchase it anywhere and no end users.

The United States government, in an effort to discourage the purchase and sale of ivory, publicly crushed tons of confiscated ivory in Denver in 2013 and New York's Times Square in 2015. In 2014, New York and New Jersey banned the ivory trade within their state borders, and California and Washington followed in 2015. In July 2015, in Nairobi, with Kenyan President Uhuru Kenyatta sharing the stage, President Obama announced a new policy "that bans the sale of virtually all ivory across state lines" within the United States.

A month prior to President Obama's action, the Chinese government destroyed 662 kilograms of confiscated ivory in Beijing—showing that the largest consuming nation in the world was starting to reverse course and get on board with the worldwide movement to end poaching. "We will strictly control ivory processing and trade until the commercial processing and sale of ivory and its products are eventually halted," Zhao Shucong, then head of the State Forestry Administration, told the media. And just two months later, President Xi Jinping of China came to Washington, D.C., to announce a shared commitment with President Obama and the United States about climate change—also a critical issue for the wildlife of Africa and the rest of the world.

But what attracted less media attention was their joint and startling statement committing the two countries to "enact nearly complete bans on ivory import and export . . . and to take significant and timely steps to halt the domestic trade of ivory." Then, as if that turnaround wasn't enough, China during the following month announced a one-year ban on the imports of trophy tusks—in this case, adopting a stronger regulatory measure than the United States, which is still allowing some limited trophy imports. Indeed, with China dramatically aligning itself with elephant protection—as the nation responsible for so much of the demand and without a longstanding ethic of concern for wildlife—it signals an extraordinary inflection point in the decades-long fight against the ivory trade. The principles of the humane economy are capable of gaining traction in the most-difficult-to-reach places and winning over people in power who never before seemed to be candidates to join our team. China years ago emerged as an economic superpower, and now it's clearly decided it no longer wants to be an outlier on climate change, ivory trading, and other environmental and animal-protection issues. And in a sign that the emerging concerns of African nations and Chinese leaders are being knitted together and put to work on the ground, Tanzania authorities in October 2015—in what is described as the biggest apprehension of a wildlife trafficker in the last decade—arrested Yang Feng Glan, the so-called Chinese "Queen of Ivory," and charged her with smuggling at least 706 elephant tusks that authorities say are worth about $2.5 million. Change is not only possible, but it's probable with diligence, diplomacy, and a rational accounting of the moral and economic features of a major problem.

All of us can live without ivory trinkets, and the dealers can trade in some other commodity and make profits anew. Kill the elephants for their ivory, however, and you not only snuff out their lives, but you choke off economic activity that provides livelihoods for millions of people. Unlike ivory as a commodity, there are no substitutes for wild elephants. They are irreplaceable: the powerful symbol of Africa, the foundation stone for the African ecotourism business, and a critical agent in maintaining the ecological health of the continent. Could the issues be any more plain, or the moral pathway more obvious for the world? We should be relentless in destroying the poaching networks, because by doing so we'll better protect not only ecological communities, but also the malls, schools, business centers, and other public places that these terrorists have targeted with dollars earned from the ivory trade. When you look at it from a moral and economic perspective, it's the future of Africa itself that is at stake. There's only one way forward, and that's to put everything we've got into stopping the monstrous, self-defeating slaughter of elephants, lions, rhinos, chimps, gorilla, and other wildlife that deserve so much better from us.

Diversifying Our Holdings in Wildlife Stocks

As we launched from the north shore of Cape Cod on a cloudless spring day, all one hundred of us were hoping for the same thing: a glimpse of a creature bigger than any dinosaur; yet whose languid movements left only a modest ripple and a gentle wake.

We paused and swayed gently in Stellwagen Bay, whose curvature and seasonal currents produce an upwelling of cold water rich with plankton and schooling fish such as herring and mackerel. It was the perfect place to lay in wait, as whales and dolphins gathered in unusual numbers. We had traded about $50 per person for a chance to smell the mist of the sea and marvel at these magnificent creatures. That day, I was a whale watcher—one of tens of thousands of people sustaining the boat owners and operators of the whale watching business.

There was a time when venturing out on a boat merely to catch a glimpse of live whales would have seemed pointless and bizarre. In the 1800s, whale oil was one of the most practical and highly valued liquids known to humankind. The "color of morning sunlight," it was sold in clear bottles and aluminum cans, stamped with a New Bedford, Mass, USA, seal. Machinists and industrialists lubricated their machines with it; the US Navy relied on sperm oil to lubricate the engines of its submarines and aircraft carriers. (Animal feed manufacturers also ground up whalebones for use in dog food.) With its clear, bright, odorless flame, sperm whale oil brought light into the homes of families across the nation and fueled the burgeoning industrial economy of the United States.

Just as news of gold in California's mountain streams spurred men to hurry west in search of riches in the mid-1800s, the lure of solid wages from whale oil attracted a generation of sailors bent on hunting expeditions that lasted up to four years at a time. In New Bedford and Chatham, giant whaling ships equipped with hundreds of empty barrels plied the world's oceans to turn whales into currency. Demand was so high that by 1854 a half-gallon of sperm whale oil cost more than $50, or about $1,400 in today's dollars. Whaling was a big piece of America's early industrial economy.

But in the late nineteenth and early twentieth centuries, we discovered fossil fuels and lubricants that are cheaper, more efficient and accessible, and far more abundant than whale oil. The innovation was timely—because whalers, with their increasingly efficient ships, sophisticated navigation tools, and exploding harpoons, had just begun the era of industrial whaling. Now, with the advent of alternatives, we were free to reexamine our relationship with whales. Over time, as we marveled at their extraordinary size and came to understand more about their intelligence, whales became a commodity we were willing to pay to watch rather than to kill, creating an economic incentive to protect them.

The global transition away from whaling is almost complete—with just three or so nations clinging to whaling as a cultural and economic enterprise, and being increasingly shunned around the world for their willfulness and cruelty. It's hard to imagine whaling ever making a comeback, or even making an enduring last stand, since there's hardly anyone interested in whale meat, and there's nobody interested in using whale oil. Very few Japanese consumers, and practically no one in Iceland or Norway, have whale meat in their diet. People have moved on from all of this. There's no market for these old ways. But it's certainly true that hundreds of millions of people would love to lay their eyes on the largest creatures who've ever lived on Earth, and many of them would pay for the privilege.

Here and there, you'll hear the argument that small-scale whaling by the Japanese and the Norwegians can coexist with the emergent economy of whale watching. But we've had a chance to consider that scenario in the most literal way. In the northern part of Norway a few years ago, tourists saw a whaling ship come into view and harpoon one of the minke whales they were admiring. "The blood flowed and it wasn't a pretty sight. This really wasn't what we came to see," one Dutch tourist told a Norwegian newspaper.

Even the people involved in whaling recognize that the world is moving away from them. Take the case of Gaston Bess, one of the last remaining harpooners in the Americas, as an indicator of where we're headed. Bess participated in twenty successful whale hunts over the course of his thirty-plus years at sea. When he retired in 2013, he declared that whaling, once a productive industry in St. Vincent in the Grenadines, "should be a thing of the past. It doesn't add anything to our economy." He now advocates replacing whaling with whale watching.

Indeed, throughout the world, political leaders are alert to the economic opportunities offered by investments in living capital. In July 2015, I sat down in New York with Rwandan President Paul Kagame, who told me that tourism generates $300 million per year for his small country, tucked between the nations of Tanzania, Uganda, and the Democratic Republic of Congo. Tall and thin, with a soft-spoken and polite manner, Kagame said that a million people come to Rwanda every year as tourists, and "more than half of them come for appreciation of nature, especially the mountain gorillas."

It's high-end tourism, strictly controlled (no more than eighty people see the gorillas in a day) in the forests of Volcanoes National Park made famous by the martyred Dian Fossey and the film that brought her life to the big screen, *Gorillas in the Mist*. It's the largest industry in

Rwanda, and President Kagame's primary task is to manage the mass of people who want to enjoy that. "They pay $750 for a permit [for non-nationals] to see them in the lush and deep green forests, and we return 5 percent of proceeds to the local community," he told me. The gorillas live in family groups of fifteen to thirty-five, he said, and "though we don't do anything to hold them in our country, they have primarily been staying in Rwanda. I came up with the idea of sharing revenues with our neighboring countries of Uganda and DRC." Mountain gorillas are the most endangered great apes in the world, with a distinct population of five hundred or so living in and around Rwanda in Virunga National Park and a smaller, distinct population living in Uganda's Bwindi National Park. With Rwanda's protective efforts and similar programs in neighboring countries, the world is losing very few mountain gorillas to poaching, and the population is gradually recovering—creating more opportunities to see them.

Elected president more than a decade ago, Kagame's enlightened approach to wildlife has not been an easy inheritance. Less than a decade before he became president, his nation experienced the closest thing to a holocaust since World War II, with the butchering of 800,000 Hutus and Tutsis in ethnic conflict that turned into genocide. "There are no devils left in Hell," a missionary said at the time. "They are all in Rwanda." In the frenzy of killing that spread across borders, some wondered whether it would be the end of gorillas too.

Since those horrific days, Rwanda has invested in protecting its national parks and wildlife from encroachment—a serious issue in a country about the size of Maryland but with twice as many people. Many of the ten million Rwandans are involved in agriculture, and that requires land, a typical formula for conflict with wildlife. Despite the presence of so many farmers, Kagame worked in July 2015 to reintroduce seven lions into Akagera National Park, nearly fifteen years after cattle herders poisoned Rwanda's last wild lions. "There were conflicts between farmers and wildlife, with the farmers retaliating by killing lions if they strayed outside of the park and onto their farms," he told me. "We built a fence to separate them and to reduce conflict, and it's worked." The distribution of revenues from wildlife tourism has given people a stake in their protection, and the fencing that has helped to keep the peace.

Other nations are also building up the business of protecting nature and creating wildlife watching opportunities. Costa Rica has safeguarded wildlife and other natural wonders as an investment in its future economic development. Nepal has preserved nearly one quarter of its lands as parks and nature reserves, and has done a remarkable job of eliminating poaching. And Kenya's national wildlife service has adopted a mission "to save the last great species and places on earth for humanity." Throughout the world there are parks, sanctuaries, and wildlife preserves—not only because these places nourish the human spirit, but also because they provide jobs and economic opportunities for communities.

Indonesia, the world's largest archipelago and its fourth most populous country, was once the biggest shark and ray fishing nation of all. But just a few years ago, the government of the Raja Ampat Regency in Indonesia's West Papua province reversed course and passed legislation forbidding the killing of all species of sharks and rays and later created a shark and manta ray sanctuary, the first of its kind in the Coral Triangle area. In early 2014 the national government doubled down on its protective efforts with legislation by declaring the entire Indonesian economic exclusive zone a manta ray sanctuary. A 2013 study examining the value of manta ray tourism versus manta ray fisheries found that a single living ray in the South Pacific was worth roughly $2 million dollars over its lifetime while a dead one sold in fish markets for only between $40 and $200. Recognizing that sustainable fisheries and marine-based tourism are the lifeblood of the nation, President Joko Widodo instructed the Indonesian Navy and Ministry of Marine Affairs and Fisheries to publicly sink any large illegal fishing vessels caught in Indonesian waters. Toward the end of 2014, the navy burned and sank six of them. In early 2015, the navy sank a large Vietnamese ship carrying two tons of drying shark fins, strips of flesh from at least five manta rays, and nearly fifty hawksbill sea turtles in its hold. The sunken ships—with their history of death and destruction—are now undersea habitats for marine life, which has returned already in abundance. Divers flock to see it.

In the Pacific Island nation of Palau, which declared its waters a sanctuary free of shark fishing, these ancient predators are the cornerstones of the scuba tourism industry. A study by the Australian Institute of Marine Science estimated that the annual value to the tourism industry of an individual reef shark that frequents these areas is $179,000 to $1.9 million over its lifetime. By contrast, a single dead reef shark is worth just $108.

Now many nations are banning the killing of sharks for their fins, along with the sale of shark fins. In China, the largest market for shark fin soup, the government halted the serving of the dish at state dinners, and the world is waking up to the plight of hundreds of species of sharks and their critical place in ecosystems. Certain species, such as the massive whale shark—which can reach forty feet in length, weigh more than twenty tons, and live more than a century—are magnets for divers and other tourists. One of the biggest gathering places for the world's largest fish is found around Isla Mujeres, a small island in the Caribbean. "The Yucatan's whale shark tourism industry has grown tremendously, increasing from just a few hundred tourists a year to more than 12,000 annually," according to the World Wildlife Fund.

Indonesia's interest in wildlife preservation extends beyond sharks. The nation is home to a number of rare creatures, among them the endangered Asian elephant and the Javan rhino. In 2014 the Indonesian Council of Ulama issued a fatwa on the "Protection of Endangered Species to Maintain the Balance of the Ecosystems," declaring the hunting and trading of endangered species to be haram (forbidden).

It had a constructive and benign message, calling on Indonesia's 200 million Muslims to take an active role in protecting and conserving endangered species, including tigers, rhinos, elephants, and orangutans. "This fatwa is issued to give an explanation, as well as guidance, to all Muslims in Indonesia on the sharia law perspective on issues related to animal conservation," said Hayu Prabowo, chair of the Council of Ulama's environment and natural resources body. "People can escape government regulation, but they cannot escape the word of God."

These aren't entirely new ideas to Indonesia, which welcomed Birute Galdikas to the island of Borneo in 1971. Galdikas, then a twenty-five-year-old UCLA graduate student, was studying under the famed British anthropologist Louis Leakey. Leakey had already sent Jane Goodall to Tanzania to study chimpanzees and Dian Fossey to Rwanda to study mountain gorillas. Galdikas persuaded Leakey to send her to Borneo to study orangutans, completing the team of pioneering women scientists whom he called his "Trimates."

But Galdikas soon realized that her mission in Borneo would have to extend beyond research. Several weeks after she arrived in Borneo, she heard from locals that loggers had captured a baby orangutan. Galdikas sent her husband out to investigate, and he soon returned with the terrified baby, whom they named Akmad. Galdikas nursed Akmad back to health at her campsite, Camp Leakey, and Akmad slowly returned to life in the wild.

Akmad was just one of many orangutans orphaned by the captive primate trade in the 1970s and 1980s. A 1986 hit Taiwanese TV show, *The Naughty Family,* starred an orangutan portrayed as an ideal family pet. In the ensuing years, an estimated two thousand baby orangutans were exported from Borneo to Taiwan alone, while another four thousand or so likely died in botched captures and transport. Some of these orangutans were doted on: one juvenile female raised in Taiwan understood many words in spoken Chinese, learned sign language, rode in cars, ate at restaurants, and even played simple pieces on the piano.

But many more were kept in tiny cages or chained in backyards. And when they grew too big to handle, nearly all were abandoned, caged, or killed.

The orangutan trade had long been illegal in Indonesia, but the government had turned a blind eye because of pressure from the booming palm oil and forestry industries. In the 1980s and 1990s, palm oil plantations spread out across Borneo, clear-cutting the peat swamp jungle in their path. Orangutans, who eat the fruit of oil palms, were seen as an obstacle to economic progress.

But as Galdikas brought global attention to Borneo's orphaned orangutans, the government's calculus began to shift. Galdikas brought National Geographic photographers and celebrities, including Sir Richard Attenborough, to Camp Leakey to learn about her efforts. The publicity brought ecotourists, who spurred the development of a local tourism industry. Slowly, the Indonesian government began to see the promise of a more humane economy, centered on admiring orangutans instead of killing them. The government sent confiscation teams to help Galdikas rescue orphan orangutans, and later set up its own rehabilitation centers.

Borneo's orangutans remain threatened, but at Camp Leakey and other rehabilitation facilities across Borneo, a more humane economy has taken hold. Every year, thousands of tourists come to see these orangutans as they adjust to life in the wild. The orangutans at Camp Leakey have relearned the joys of being free—swinging from trees, playing, and raising a new generation of wild-born infants. Some formerly captive orangutans will even hang hammocks up to lie in or take canoes out for a paddle down the river. Once their persecutors, we can now be their protectors, creating economic enterprises to allow people to safely interact with these great apes and committing ourselves to saving them for their own sake and for future generations to behold.

CONCLUSION

High Yield Bonds

"The reasonable man adapts himself to the world: the unreasonable one persists in trying to adapt the world to himself. Therefore all progress depends on the unreasonable man."

—GEORGE BERNARD SHAW, *Maxims for Revolutionists* (1903)

Perfect Information, Better Outcomes

"I believe that people and economies do best when they work in harmony with nature, as opposed to treating it solely as a source of resources to be extracted," Microsoft co-founder and philanthropist Paul Allen explained to me in an exclusive interview. His interest in animals turned into a deeper resolve after a series of trips to Africa, where he saw the great herds of African wildlife and the full roster of predators. The billionaire owns tourist lodges in Botswana and told me that "nobody can visit the Okavango Delta without being awestruck by the diversity and richness of its animal life. Watching a family of elephants walk single file, or listening to submerged hippos belch, or hearing a lion's coughing roar cut through the canvas of your tent all have a way of turning wildlife conservation from an abstract challenge into something more real and personal." He's funding wildlife scientists on the ground, and he's also financing the Great Elephant Census, a pan-African collaboration that will generate the first accurate count of savannah elephants, their migration patterns, and poaching hotspots. "Once we get this data, we're going to become a lot more effective in protecting them," he tells me.

He's also a scuba diver and similarly awed by marine life. And he blended all of those passions into the most comprehensive anti-wildlife trafficking ballot measure in any US state, since it is the United States and China that have been the biggest consumers of wildlife parts. In November 2015, Washington voters, after a campaign led by Allen and the HSUS, put a stop to the trade in parts of elephants, rhinos, lions, leopards, tigers, cheetahs, shark, and rays. Even though the National Rifle Association and some antique shop owners fought the measure, it passed in every county in the state, with more than 70 percent of the vote. Poachers and wildlife traffickers are outside of the humane economy, and that's where government strictures come into play and create a new bright-line standard for society.

The bond with our fellow creatures runs deep for so many of us—American and African, rich and poor, young and old, male and female. It's not a peculiarity or a quirk or sentimentalism run amok. It's being alert to the feelings and needs of other creatures, and making adjustments in our lives to reflect that awareness. No, there's no prescribed set of experiences that kindle the feeling and no formula for awakening others to it. Animals do not touch every heart in the same way. But every one of us—including many of the richest and most powerful people on the planet—has had small and large experiences that have kindled something in them and connected them to animals, inspiring acts of kindness and even leading some among them to challenge old customs or archaic practices.

The bond stirred in me in my earliest years, especially on the athletic fields across the street from my home where I'd throw the ball for hours on end to Brandy, my retriever mix. I eagerly awaited every issue of *National Geographic* and each episode of *Wild Kingdom*, marveling that the blue whale is nearly one hundred feet long or that packs of wolves live together as families and maintain a strong social hierarchy. After learning about destructive fishing practices, I appealed to a power even greater than City Hall, asking my mother if she'd refrain from buying any tuna caught by setting nets on dolphins.

Our human connection to animals finds so many surprising expressions, revealing itself even in the most trying and dire of human experiences. It had to be one of the oddest-looking hostage releases in history. In April 2012, the Revolutionary Forces of Colombia (FARC), perhaps the most notorious leftist revolutionary group in South America, had freed its last remaining hostages. A few of these former police officers and soldiers had been captives for fourteen years in a jungle hideout, separated from family, friends, and everything that mattered to them. Their freedom came after a breakthrough in negotiations between the FARC, government leaders, and the International Committee of the Red Cross. A Brazilian military helicopter on loan to the Red Cross had transported them from their hideaway to an airport in the city of Villavicencio, before heading on to Bogota. As the hostages stepped from the transport helicopter and onto the tarmac, a film crew recorded not just their jubilant reunion with their families, but also an unlikely processional.

At the heel of police sergeant José Libardo Forero, one of the freed men, there was a quick-stepping, sinewy wild pig, a peccary, keeping pace. The peccary, who probably weighed forty or fifty pounds, wasn't on a leash, but he never ranged more than a foot away from his human companion. Forero said he took in the animal when he was only two days old, naming him Josepo and feeding him milk with a syringe. A captive for thirteen years, Forero pointed to a scar on his forehead where Josepo bit him. "They are aggressive," Forero said. "They are savage. But I trained him and he doesn't bite anymore. . . . I took care of him."

A second freed hostage, dressed in his army fatigues, walked out with two green parrots on his shoulder—wild animals he befriended during his time as a captive. And Olga Lucia Rojas cradled a raccoon-like kinkajou in her arms as she spoke with a CNN affiliate. She said her brother, former hostage Wilson Rojas, brought the rainforest animal home with him. "His name is Rango," she explained. "He adopted it in the jungle. He told us it was his companion. They wouldn't separate for any reason."

The Colombian soldiers and police, denied by their captors of almost any freedoms or conveniences, had bonded with animals not to pass time, but to survive the torment of isolation. Their stories remind us that pets and other animals not only enrich our lives; sometimes there's a void that only they can fill, and comfort and happiness that only they can bring. We have a bond with them, seen on that tarmac in Colombia, or in the greetings that animals give us after we come home and push open the front door, or in the thrill we feel when we catch a glimpse of an elk or a dolphin in the wild.

Not to rob this bond of all its mystery, there may even be some biochemistry at work that helps explain it; just as there is in the bonds we form with one another. The hormone oxytocin, which promotes maternal care, friendship, and romance between humans, also appears to play a pro-social role in our interactions with animals. Time spent with animals, or even the sight of one, can release a rush of oxytocin, which can in turn slow or speed up our heart rate, calm or excite us, or generate a smile or tears. At some senior centers, administrators arrange for pets to visit with the elderly as an antidote to loneliness, and as eyes meet, there is much tail wagging and smiling. For military veterans coping with posttraumatic stress disorder, dogs and other animals are a form of therapy, relaxing them and even giving them the confidence to face the world.

Yet this natural empathy and connection with other animals is not the final word, and it leads to no predetermined outcomes. There are competing instincts, whether greed, desire, or more atavistic and specific impulses like hunting and killing that can trump our inclination toward more compassionate and nurturing ways. We can be pulled in opposite directions, and it may be social pressure or conditioning or reason that pushes us down one path or the other.

One challenging problem for us, in modern society, is that we can be so disassociated from the reality of exploitation, removing the urgency and even the moral influence or relevance of an issue. Most people don't buy ivory or rhino horn, set a snare or a body-gripping trap for an animal, confine a pig in a crate, drip chemicals into the eye of a rabbit, and intentionally hurt or kill an animal in some other way for profit. A kind-hearted person couldn't imagine doing such things. Yet with the use of animals built into the routines of so many businesses, and with supply chains that stretch for thousands of miles and, at their point of origin, are hidden behind fences or laboratory doors, somebody else does that work. They are our proxies. And for that reason, just about every one of us is implicated in those very practices we would never be a party to if we saw them up close.

Unless you are consciously avoiding them, you may buy parkas or gloves trimmed with rabbit fur. You may buy cosmetics or household products tested on animals. Or you may buy meat and other animal-based foods from factory farms. Perhaps you know in the back of your mind that there is more than a strong residue of cruelty in such products, but almost never focus on the painful details. You become an unwitting or passive commercial partner in these enterprises, hardly realizing the intersection of your purchasing practices and the moral problem of cruelty. Similarly, the people actually causing the harm directly don't want us to be reminded of their practices—in fact, they often make very determined efforts to conceal the cruelty.

In considering this question, we need to broaden our understanding of moral responsibility to animals. Cruelty isn't limited to practices that we in society hold in general contempt, such as beating a dog or setting a cat on fire, or even long-established but now widely reviled activities such as dogfighting or commercial whaling. The abuse of animals is also embedded in economic enterprises that provide staples in our daily lives. Because it's both routine and invisible, and we play no direct role in the violence, we generally consider it a moral problem not of our making. Our physical removal from harmful actions toward the animals provides a sort of buffer. Yet as end users, we enable it, just as we have the power to deny it by abstention. It is only consumers who give these products value, or, conversely, who can make them worthless in the marketplace.

Some of these practices are so deeply ingrained in our culture it's hard for us to see them clearly, even as they stare us in the face. Timothy Pachirat, a political ethnographer, spent six months working undercover on the floor of a cattle slaughter plant in Omaha, Nebraska—a facility that killed and disassembled 2,500 animals a day, transforming three-quarter-ton animals into hundreds of packaged meat items sent to stores and restaurants throughout the nation. On one eventful day, a group of cows escaped a pre-slaughter pen and made a dash for it—into the streets of Omaha. Four of them ran for the parking lot of the St. Francis of Assisi Catholic Church. But none of the disciples of the patron saint of the animals came to their aid, and the four were rounded up and returned to their holding pens to await slaughter. Another cow, however, didn't cooperate, almost certainly realizing that capture meant death. After a while, "the police waved the workers back and opened fire," emptying a dozen rounds from their firearms and fatally wounding the animal, who bellowed pitifully before expiring.

Pachirat reports that the workers at the plant were outraged at the police killing of this cow, talking in their lunchroom about how callous and unnecessary it was. Remember, these people were the line staff in a business responsible for killing ten thousand cattle every week. If that cow had not found an escape route that day, she would have been shot in the head with a captive bolt gun, hung up by a back leg, and taken apart quite unceremoniously, unremembered in the blur of a hundred kills an hour.

During his employment at the plant, Pachirat diagrammed the floor plan and identified the duties of the 120 people on the slaughterhouse floor. He concluded each line worker had a narrow and specific task—for example, as a tail ripper, side puller, or paunch opener, in the lexicon of the industry. Only 4 of the 120 were actually involved in killing animals, and just about everybody else had a role of removing, or setting up the removal, of small parts of the corpses. Their work was so compartmentalized, observed Pachirat, that they lost perspective on the larger enterprise and their own contributions to killing animals. This partly explains their shared indignation at the shooting of the runaway cow.

If the people on a slaughter plant floor can feel disassociated from killing, consider the larger dynamic that has taken hold in our society. The employees worked anywhere from a few feet to a couple of hundred feet from the kill box. The rest of us are sometimes removed from it by hundreds or thousands of miles. Most of us have very little interaction with the people involved in killing animals in these sorts of enterprises and only partial knowledge or murky understanding of the details. The meat we buy in stores comes neatly packaged or well presented on a restaurant plate. We know it's an animal, but we've become so accustomed to the product, and so removed from the process of how it got there, that it becomes perfectly reasonable to accept the whole enterprise. When the fast-food company KFC (a.k.a. Kentucky Fried Chicken) allowed a BBC film crew into a slaughterhouse that supplies the chain with chicken for a series, pundits noted what a "very big gamble" it was for KFC. Mark Lang, a food marketing specialist at Philadelphia's St. Joseph's University, told Bloomberg News that his research shows that most people would rather not know the details of where their meat comes from: "They don't want to be reminded that it was a live animal."

Indeed, such reports and investigations, including those conducted by HSUS and other groups like Mercy for Animals or Compassion Over Killing, are closing the gap between what we're told to believe and what actually happens to animals. But it doesn't come easily or without resistance. There's undeniable momentum behind the humane economy, but animal-use groups continue to fight rearguard actions in defense of the status quo. The most in-your-face example of this is the effort by industry-backed lawmakers in more than half a dozen major agricultural states to pass so-called "ag-gag" measures. These laws make it a crime to take pictures of animals on factory farms, or for animal advocates to apply for jobs at these facilities. Several states have amended their state constitutions to add right-to-farm measures that enshrine existing agricultural practices and ward off future attempts to restrict them. Even more widespread than these laws against whistleblowers and for categorical protections for animal-use industries are government subsidies to favor the meat industry—providing price supports for feed crops, crop insurance, predator control programs, research and development funds, and programs to purchase surplus meat products, which are then fed to prisoners or school kids. This crony capitalism, in which the state distorts markets for private interests and corporations, is at work across the animal-use economy, but it's most extreme in industrialized agriculture and in the field of biomedical research, and it retards reform and elimination of archaic and cruel practices.

But these industries err in assuming that consumers have static views, or that they will continue to accept the imperfect information and outcomes they deliver. The humane economy, when it's working, is capitalism at its best, engaging the problem of cruelty to animals and applying human creativity to answer the demands of a morally informed market. To borrow a popular word in business these days, the humane economy is disruptive, and its influence is being felt by corporate livestock producers (do they even deserve to be called "farmers" anymore?), trappers and furriers, animal experimentation laboratories, puppy mill owners, circus operators, and the whole range of industries that trade on the misery of animals. For the most part, they fight change at every turn, reacting with suspicion and indignation to even the least suggestion that their practices are morally untenable. Yet for most everyone else, the humane economy is a series of changes and choices that add up to a welcome moral opportunity, a chance to make our own lives better, and for all humanity to leave a kinder mark in the world.

On some days, my more optimistic ones, I believe the time will come when all the factory farms, fur ranches, and other places of systematic cruelty are gone, except for the few that may survive here and there as museums or memorials to preserve the evidence of what life was like when creatures were confined and tormented by the billions, by people who congratulated themselves for the pace and efficiency of production. When such places do one day disappear, who will miss them? If the world had moved on to good and nourishing meat substitutes—a conversion already well underway—who will make the case that things were better in former times? Who will look back with nostalgia on the "mass confinement" facilities of today; the around-the-clock, workmanlike dismembering and killing; and all that goes with the industrial farm economy as we know it?

In the literature of the discipline, economics can often seem as little more than a grim grinding of the machine, powered by the coldest, most impersonal of forces—bidding, in the words of Dickens, "all human sympathy to keep its distance." It's an impression on which backward-facing industries depend, because it keeps things just as they are and makes change seem hopeless. And yet it's a false impression, obscuring the underlying truth of the free market—the collective power that belongs to every consumer. Indeed, consumer choice informed by conscience is an unstoppable force for good. Today, in fact, there's more progress, more communication, more commentary, and more action on animal issues than ever before—a veritable revolution in information, thought, and practice. The voices of inertia once dominated the debate, but there is now an inversion happening among opinion leaders and it's a dangerous time for defenders of the status quo.

Charles Krauthammer, the nationally syndicated conservative columnist, argued in a May 2015 column that "our treatment of animals" will be deemed "abominable by succeeding generations." He observed that "our great-grandchildren will find it difficult to believe that we actually raised, herded, and slaughtered [animals] on an industrial scale—for the eating." The conservative columnist Kathleen Parker followed a similar line of thought, asking, "Is making a hen's cage a little larger really so cost-prohibitive that we can't manage to make a miserable life a tiny bit less miserable? Is someone's taste for foie gras so worthy of protection that we condone force-feeding cruelly confined ducks until their livers bloat and become diseased?" These days, it's hard to find any mainstream voices on the other side of major animal-protection issues. "We as a global society have crossed the Rubicon," liberal columnist Nicholas Kristof wrote in an op-ed for *The New York Times*. "We disagree about where to draw the line to protect animal rights, but almost everyone now agrees that there is a line to be drawn."

These writers remind us of the attributes of animals and the responsibilities of men and women to them. For too long, industry had the thought leaders, lawmakers, and retailers and other business leaders on its side, and together they belittled a social concern for animals, brushed aside the practicality of reform and the availability of alternatives, and denied even the capacity of animals to suffer. They controlled the scientists and the language and the framing. You see this in so many trade journals still. Wild animals are "game" to be "harvested on a sustained yield basis," lab animals are "tools for research," and farm animals are "units of production." Language has always been used by those in power to characterize animals as insensate—as if they're just populations and not individuals, who don't share familiar characteristics or attributes and are born for exploitation. When you are in denial, and cast the victims as unworthy of empathy or compassion, it clears the path for exploitation. This is a familiar linguistic pattern in some of history's most barbaric or appalling cases of human exploitation and violence. The historian David Brion Davis wrote of the "systematic animalizing of African Americans" in the antebellum South. According to Theodore Dwight Weld, whose work was published by the American Anti-Slavery Society in 1839, "Slaveholders regard their slaves not as human beings, but as mere working animals, as merchandise."

I was struck too by the language used by the designer Karl Lagerfeld in a 2014 *New York Times* interview. Now in his eighties, Lagerfeld described his great affection for his cat, Choupette, who has more than forty-eight thousand followers on the Twitter account he created in her honor. Lagerfeld said he hoped the cat would become more famous than him, declaring "then I can disappear behind Choupette." The occasion for the interview was Lagerfeld's fiftieth year with his fashion label Fendi, and his latest line of clothing, as in prior years, included an abundance of animal fur. A reporter asked if he was a critic of the anti-fur cause. But Lagerfeld said he's "very sympathetic" and "hate[s] the idea of killing animals in a horrible way, but I think all that improved a lot. I think a butcher shop is even worse. It's like visiting a murder. It's horrible, no? So I prefer not to know it."

So here's a man who loves his own fur-bearing cat—in such an affectionate and heartfelt way. But he's using fur from other animals, who don't have fancy French names but are in every way the moral equals of Choupette. He professes confidence that animals in the fur trade are killed more humanely, but apparently is unaware that there are no government regulations or any meaningful industry standards on animal welfare for that industry. And he then engages in a classic bit of redirection, by noting that other forms of cruelty, such as animal slaughter for food, are even worse. Ultimately, Lagerfeld concedes he doesn't know anything about it and doesn't want to know because it's too painful to contemplate. He's just the designer—let somebody else worry about the rest.

Happily, while he lives with that kind of moral confusion, other designers see things more clearly and are moving rapidly into the future. In July 2015, the luxury clothing manufacturer Hugo Boss ended any use of fur, saying "we will not be using any farmed fur" and the company notes that "acting sustainably sometimes means saying 'No.' " Giorgio Armani let us know a month later that his company is doing the same, with fur out of his lines by 2016. When we can produce faux fur that is indistinguishable from real fur, why would we trap wild animals or raise them on fur farms, subjecting them to torment and death? The designer and the seller simply work with a different product, and in the end, the consumer is not asked to make a sacrifice or experience a hardship—only to go to a different rack in the store and buy a functionally equivalent product not woven with moral problems. The faux fur maker buys fabric and provides jobs for manufacturing, and the retailer still sells the product to generate profits and employ sales staff. There are just no animals killed by cruel means. It's an example of creative destruction—but with the creativity triumphing and the destruction nowhere to be seen.

It's a simple equation, really. We free animals from abuse, and we free ourselves from reliance on abuse. These days, to be an unquestioning customer of industrial meat products or the fur salon requires carefully tended habits of euphemism and denial. Who really wants to go on like that, when better and worthier options are right in front of us?

Clinching the Case, Stepping into the Future

The election in 2013 of Argentinian Cardinal Jorge Mario Bergoglio to lead the one-billion-member Roman Catholic Church was met with widespread praise, given his reputation for humility and selfless concern for the poor. At the HSUS, we were abuzz with excitement when we learned the new pontiff had selected his name in tribute to St. Francis of Assisi, the patron saint of the poor as well as animals and the environment. From the start of his papacy, he's not disappointed those who had high hopes for reform coming out of Vatican City, and no public declaration was more forceful or eagerly anticipated than Francis' two-hundred-page encyclical about animals and nature, issued in June 2015.

"Our indifference or cruelty towards fellow creatures of this world sooner or later affects the treatment we mete out to other human beings," wrote Pope Francis in *Laudato Si* (*Praised Be*). "We have only one heart, and the same wretchedness which leads us to mistreat an animal will not be long in showing itself in our relationships with other people. Every act of cruelty towards any creature is 'contrary to human dignity.' "

These are powerful, unambiguous statements from the world's most influential religious leader, who called on the global community to end cosmetic testing on animals, halt the abuse of animals in agribusiness, and end the extinction crisis. He associated himself squarely with the idea of the humane economy and the ethical marketplace, repeating the dictum of his predecessor, also a friend to animals, Pope Benedict: "Purchasing is always a moral—and not simply economic—act."

"We read in the Gospel that Jesus says of the birds of the air that 'not one of them is forgotten before God,' (Lk 12:6)," Pope Francis wrote. He continued:

> "How then can we possibly mistreat them or cause them harm? I ask all Christians to recognize and to live fully this dimension of their conversion. May the power and the light of the grace we have received also be evident in our relationship to other creatures and to the world around us. In this way, we will help nurture that sublime fraternity with all creation which Saint Francis of Assisi so radiantly embodied."

The Pope was not inventing a Catholic catechism on concern for other creatures—he was reminding us of it, and underscoring the urgency and importance of the task. Throughout the Bible, there are calls for mercy—the "righteous man regardeth his beast, but the tender mercies of the wicked are cruel." In fact, God in the Old Testament says quite clearly that animals belong to him, meaning they are not ours to abuse: "And the Lord said: 'For every wild animal of the forest is mine, the cattle on a thousand hills. I know all the birds of the air, and all that moves in the field is mine.' " (Psalm 50: 10–11) Old fasting practices still observed, dietary and slaughter rules, and prayers before meals keep these principles alive with Christians, while acknowledging the stain of violence against animals.

The spirit of mercy is found in all the world's great religions, and they remind us of our charge in caring for Creation. Hinduism teaches that the Divine exists in all living beings, so all animals are to be accorded respect and compassion, and many Hindu gods take animal forms, including the elephant god Ganesha and the monkey god Hanuman. Jainism calls on its followers to adhere to the principle of *ahimsa,* or nonviolence, and both Hindus and Jains have a special devotion to cows, who are viewed as symbols of motherhood. Not only are there religious and legal proscriptions on killing them, but throughout India there are thousands of shelters, known as *goshalas,* for homeless or ailing cows. Indian President Pranab Mukherjee approved a bill in March 2015 that strictly bans the slaughter of cows, along with the sale, consumption, or even possession of beef in the state of Maharashtra, where Mumbai is located. The highest Buddhist virtue is compassion, meaning avoiding causing suffering or death to any sentient being, and the Dalai Lama has called for an end to all forms of animal abuses. The Jewish principle, *Tza'ar ba'alei chayim,* or "the suffering of living creatures," prohibits inflicting unnecessary pain on animals. The Qur'an and Hadith of Islam both describe the Prophet Muhammad as having great compassion for animals.

Today, the invocation to be good to animals can be heard from so many directions—from the world's wealthiest men and women, thought leaders, the executives of the biggest corporations, and past and present clerics and other religious leaders. But to clinch the case for animal protection, it's necessary to look to the realm of economics. We've often heard from apologists for animal cruelty and the status quo that the use of animals is simply a harsh reality, and that it must persist in spite of sentimentalism because its commerce is needed as a means of generating jobs, profits, and growth. In so many ways, this is economic sophistry, a reflexive ploy the late British economist Arthur Cecil Pigou would have had a field day with were he alive today. Nearly a century ago in *The Economics of Welfare*, Pigou argued that private industries often impose costs on other people, with the costs not measured directly at the point of purchase. Pigou noted that there must be a full and proper accounting and a tabulating of these externalities in order to determine the full and actual costs of the operations of an industry.

Factory farming is a classic case in point. By delivering cheap meat to consumers, it may seem economical for consumers, but when attorney David Simon tried to put a price tag on the enterprise in *Meatonomics*, his rough estimate was that it costs society more than $400 billion per year. High rates of meat consumption produce hardened arteries, stroke, heart attacks, and several varieties of cancer. These diseases afflict hundreds of thousands of people every year, and that's precisely why the American Heart Association, the American Cancer Society, and so many other health-oriented authorities urge people to eat less meat and more plant-based foods. The healthcare costs associated with heart disease and cancer are in the hundreds of billions a year, to say nothing of the emotional costs to family or the impacts on businesses when members of their workforce are temporarily or permanently lost. Then there are the costs associated with pollution and government subsidies to the factory farm lobby. Sure, the cattle, pork, and poultry industries are big business, themselves, collectively generating hundreds of billions in sales, and creating business for feed suppliers, truckers and other transporters, drug makers, veterinarians, food retailers, and lots of others connected to food (if people didn't eat the animal products from those industries, they'd have to eat something else, of course, and it's important to remember that other agricultural enterprises would fill the void effortlessly and also produce many billions in commerce). But when there is a full accounting of costs and benefits of industrial animal agriculture, it's not quite the bargain it's made out to be, with hundreds of billions in costs and liabilities offsetting the profits. Consumers bear these costs quite directly, in the form of higher healthcare costs, higher taxes, environmental mitigation, and property value impacts. They just don't show up at the checkout counter or on the restaurant bill. But that doesn't mean society, as a matter of economic transparency, shouldn't provide that kind of detailed accounting.

More than ever, we are coming to grips with the public health costs of our industrial system of agriculture, and we're moving toward a more accurate gauge of the consequences of a sick food system. Every year, one in six Americans—for a total of forty-eight million cases—contracts foodborne illnesses. According to a report from the University of Florida, factory farm operations cause a majority of food safety problems, including Campylobacter and Salmonella bacteria in poultry, Toxoplasma gondii in pork, and Listeria bacteria in deli meats and dairy. To this list, add bird flu, Mad Cow Disease, and Methicillin-resistant Staphylococcus aureus (MRSA), a life-threatening bacterial infection that is resistant to some of the most commonly used antibiotics in human medicine. The presence and virulence of these pathogens are often caused by unsound conditions endured by animals on the farm, including severe overcrowding and animals living above their own waste.

The threat of these illnesses is made worse by the overuse of antibiotics on factory farms. According to the US Centers for Disease Control and Prevention (CDC), in the United States, "at least 2 million people become infected with bacteria that are resistant to antibiotics, and at least 23,000 people die each year as a direct result of these infections." The CDC reports that 80 percent of all antibiotics are used on farm animals, noting that "much of the use of antibiotics in animals is unnecessary and inappropriate and makes everyone less safe." Precisely because the farmers know the animals are overcrowded and stressed, packed in so tightly, they use antibiotics as a prophylactic in order to prevent disease—a strategy that is unheard of in the field of human health care when it comes to antibiotics. Medical groups have long been concerned about overuse because it eventually renders antibiotics useless, as microbes develop resistance to clinical doses. In response to this crisis, almost every major public health organization—from the American Medical Association to the American Public Health Association—has urged a ban on the use of antibiotics for nontherapeutic purposes. In a 2014 report, the World Health Organization called antibiotic resistance "a problem so serious it threatens the achievements of modern medicine" and pegged the cost of antibiotic-resistant infections at $21 billion to $34 billion in the United States alone. Industrial-style pig farmers around the world use, on average, nearly four times as many antibiotics as cattle ranchers do per pound of meat, according to a 2015 study published in the Proceedings of the National Academy of Sciences. In 2010, the world used about 63,000 tons of antibiotics each year to raise cows, chickens, and pigs, the study estimated.

Despite the serious threats to public health, however, an axis of farm industry trade associations, pharmaceutical companies, and veterinary groups (paid to administer the drugs) have been able to fend off reform in Congress. But change has a way of finding alternate channels. A number of major food retailers, notably McDonald's, Costco, and Walmart, have taken preliminary steps to move away from buying animal products from operations that lace animal feed and water with antibiotics, but the new policies have loopholes that allow producers to keep using some classes of antibiotics for nontherapeutic reasons.

At the US Meat Animal Research Center (USMARC), the federally operated farm research laboratory, animal experimenters have been trying to increase the number of cows who birth twins, increase litter sizes of pigs, and breed "easy care" lambs. Experimenters locked pigs in steam chambers until they died, and left thousands of sheep to starve or be pelted by hail in experiments. The Nebraska center was also raising thousands of animals for slaughter—essentially running a factory farm of its own—and using the profits to do more unregulated research. This facility burned through $200 million between 2006 and 2013, and it's just one of fifty-one such research centers operated by the Agricultural Research Service of the USDA. These centers comprise another channel of government support for animal-related businesses, acting as the research and development arm for private industry and distorting the market with public money.

The economist Pigou would have found plenty more fodder in other domains of contemporary animal use to illustrate the argument that animal-based businesses are unfairly transferring costs of their enterprises to an unsuspecting public. Take the case of the closely related worlds of commercial dog breeding and animal sheltering in the nation. The bulk of the billions that private citizens invest in animal-welfare charities is spent on housing, sheltering, and caring for neglected animals at local facilities. The industries and the people that leave behind homeless and abused animals typically pay nothing for their long-term care. Puppy mills, animal laboratories, exotic animal breeders, and Hollywood animal trainers profit from selling and using animals, but then often dump their animals on shelters and sanctuaries when their work is done. Think of the work of the thousands of animal shelters filled up with dogs and cats, the hundreds of big cat sanctuaries with tigers and lions discarded by naïve owners, or even the few sanctuaries caring for chimps retired from acting when they were seven years old but destined to live for another fifty years. The costs for housing and placing homeless dogs and cats—most of whom have been bred for the pet trade and then left behind at some point by an animal owner—is in the billions just for the private charities. Add an additional billion dollars for the municipal or county animal care and control operations and you start to get a truer accounting of the real costs.

Why must animal-protection groups and taxpayers pick up the tab for businesses and people that make reckless decisions and simply transfer the costs to the rest of us? We need a set of policies to address these issues so that the problem makers pick up the bill for the cleanup and rescue rather than the well-meaning people who must do so now.

A January 2012 report by researchers with the National Academy of Sciences found that in a little more than a decade of colonizing the Florida Everglades, Burmese pythons have wiped out 99 percent of raccoons, opossums, and other small and medium-sized mammals, and 87 percent of bobcats. Some experts estimated that tens of thousands of these exotic constricting snakes colonized south Florida—the result, in all likelihood, of their cycling through the pet trade. In response to a burden they didn't want to meet any longer, some people decided to dump their animals into forests and swamps, including Everglades National Park. The snakes then went to work, showing off their enormous reproductive potential and voracious appetite. In a widely viewed video, a python and an alligator get into a swamp fight; the python kills and consumes the alligator, and the snake bursts open after swallowing the giant beast, in what can only be considered a pyrrhic victory. We as a society—and most of all, the native species crushed and suffocated by the newcomers—are stuck having to deal with the consequences of an introduced species like this. The ripple effects are felt up and down the food chain, and may even threaten the survival of the highly endangered Florida panther. We as a nation have invested billions in protecting the Everglades, and a foolish interest in having these animals as pets on the part of a few has threatened to ruin the plans of so many.

The estimated annual cost to the United States of removing and managing invasive species of animals and plants is $120 billion, or about one-fifth the cost of maintaining all of our armed forces. The federal government has banned the trade in more than a half dozen species of large constricting snakes, though only after being pressed by the HSUS. These costly misadventures should serve as cautionary examples and prompt policy makers to adopt rules to prevent additional exotic species from colonizing natural habitats. Stopping the trade in live wild animals as pets, shutting down the exotic hunting ranches, and paying attention to where the animals rightly belong are part of the answer.

Then there are the horse track owners and racehorse owners who aspire to own a Triple Crown winner and breed animals but then decide not to provide lifetime care. They either funnel horses into the slaughter pipeline, or they dump their animals onto sanctuaries—with an estimated five hundred sanctuaries costing America's animal-protection charities and donors tens of millions of dollars. If the horses are rerouted into the slaughter industry, they are purchased by "kill buyers." These people jam the horses into overcrowded trucks headed on a thousand-mile journey to a slaughter plant in Canada or Mexico. In Mississippi, authorities discovered over 150 dead and emaciated horses on the property of one kill buyer, who also had a criminal history of livestock larceny. And in South Carolina, authorities found dead and emaciated horses on the properties of two alleged kill buyers. One of them was found out only after a video was posted online showing him beating his girlfriend's dog. Kill buyer Dorian Ayache got into two accidents while transporting horses to slaughter, endangering the horses and other drivers. He reloaded the surviving and injured horses onto his trailer and tried to take them across the border to Mexico, where authorities rejected several of the animals because of their severe injuries. Ayache's operation was shut down by federal authorities, but he continued transporting horses under another company name. The horse slaughter business is not a reputable one, and it's the playing field of unscrupulous people who create havoc and harm to animals who deserved a much better fate. Evidence of these abuses finally compelled the Congress to prevent any reopening of horse slaughter plants in the United States, and now the European Commission has ended imports of horses slaughtered in Mexico, which had been the primary destination for American horses funneled into the slaughter pipeline.

Investing in the prevention of cruelty—by adopting anti-cruelty standards in the law and then enforcing those laws—will also reduce crime and violence in our society and the costs of that conduct. In 75 percent of cases where there is animal cruelty at a home, there are other forms of domestic violence too. In Chicago, among a group of people charged with animal cruelty crimes, 70 percent had prior felonies. At major cockfighting arenas in East Tennessee, federal agents broke up an operation where local sheriffs had been corrupted and the cockfighters were running illegal gambling operations, bringing children to the fights, and running prostitution rings and chop shops on the side as well. It's not uncommon to learn of shoot-outs or murders at dogfights and cockfights; it's no surprise that people who like watching animals die in a pit would settle their own scores with a burst of violence.

Many of the most malicious forms of cruelty are now illegal, thanks to our growing consciousness about animals and the terrible things done to them. Yet in other cases, exploitation and abuses occur with the encouragement or even the collusion of the government. Greyhound racing is in an economic death spiral in the United States due to almost nonexistent consumer demand and the determined work of several humane organizations, including Grey2K USA, the leading greyhound protection organization. The practice is illegal in thirty-nine states, and only nineteen tracks in six states operate—down from forty-nine tracks in fifteen states in 2001. The handle—the amount wagered on greyhound racing, live racing, and simulcasting—has declined from $3.5 billion in 2001 to $600 million today. The few states running dogs have a statutory live racing requirement if other forms of gambling occur at the tracks—obligating reluctant casino owners to keep alive a spectator sport that just a handful of people watch, that loses money, and that harms dogs. "Dogs are running around in circles with nobody watching," according to Carey Theil of Grey2K USA, characterizing a typical night at the track.

It's not quite as grim an economic forecast for the horse racing industry, but there are similarities. Many lower-tier horse tracks—their grandstands almost empty of racing enthusiasts—survive only because their enterprises now include elegant gambling parlors, with slot machines and card rooms, which are often physically separated from the horse-racing oval and that whole somber, desolate scene. The kicker is, the states stipulate that if the tracks are to operate casino-style forms of gambling, they must also run horses, even as the racing side of the business is hemorrhaging money and losing millions. In the end, the card games and slots are paying to keep the lights on and to prop up the horse racing industry, and here again, we see the role of the state in distorting the marketplace, and in this case, picking a loser and forcing someone else to bear the cost; the state laws requiring horse racing at the tracks are keeping these backwater tracks alive. There's also widespread evidence of doping of racing horses—to enhance performance and oftentimes simply to get injured or lame horses in the starting gate. The *New York Times* reports that twenty-two horses die on racetracks each week in a sport that's supposed to celebrate the vitality and athleticism of the animals, and not to turn tracks into crash sites for animals that break down.

In Canada, the business of killing baby seals to provide their fur has resulted in markets throughout the world—the European Union, Mexico, Russia, and the United States—saying they want nothing to do with this kind of commerce. For the moment, a much-reduced hunt limps along only because the Canadian government buys up the pelts as a subsidy for a hidebound industry. The government keeps dumping money into the hunt because local residents in coastal villages in Newfoundland or Prince Edward Island support it, but if they don't use the pelts locally and no one wants the pelts anywhere in the world, what is the point? Why not just hand over the money and spare the seals? Why provide subsidies for a moribund industry when the government can help with job creation and other forms of economic development for an evolving economy?

Consumers also often assume that industry and government are doing right by animals, trusting them to act with integrity—a belief that industry at least works hard to cultivate. In recent years, Perdue Farms affixed "humanely raised" to labels for its chicken even though it is widely recognized that the company relies on unnaturally fast-growing birds who live in chronic pain, and utilizes transport and slaughter practices that are demonstrably inhumane. The HSUS filed a legal action to stop the company from advertising that false claim, and Perdue agreed to remove the label to settle the case. But Perdue continues to advertise and sell the notion that the company's birds are raised in "cage-free" environments, which, while literally true, rings hollow because no "meat birds" are kept in cages—largely because birds living in cages become so badly bruised that consumers won't buy the meat. Only laying hens are kept in cages, but since Perdue isn't in that business, it's knowingly making a meaningless declaration.

Yet a correction seems inevitable because of greater access to information and more organizations devoted to calling out deception or cruelty when they see it. Of Americans with pets, 91 percent consider the pets to be family members, according to a 2015 Harris poll. After government authorities issued mandatory evacuation orders prior to Hurricane Katrina's making landfall in 2005, thousands of Gulf Coast residents decided to hunker down at home because no shelter would allow them to bring their pets. And after the storm hit and the levees collapsed, flooding streets and disabling utilities, thousands of people turned away rescuers because emergency responders had been given misguided instructions to take "only people, no pets." No one who has watched the nightly news should have been surprised, for it's not infrequent to hear of cases where some Good Samaritan plunges into a frozen lake or rushes into a burning home to pull an animal from a life-threatening situation. Most people view these acts as heroic, and who among us doesn't hope that we'd muster that kind of courage in a similar crisis? Between our pro-social hormones and an ever-more-widely-held and popular value system that calls on us to defend animals, we've got a head start in doing the right thing. But it's up to each one of us, as creatures of conscience, to make life-affirming choices. More and more people are making compassionate choices, and it's a reflection of the changing values and opportunities in our society.

The arc of human experience tracks ideas that seemed impossible until they become universal. Look at the progression of changes that have altered the fundamentals of human society: in composition, we've gone from the stone tablet to the printing press to the typewriter to the digital tablet; in information exchange, we've shifted from the telegraph to the telephone to the fax machine to email and the Internet; in transportation, we've moved from the horse to the streetcar to automobiles and trucks to airplanes; and in photography, we've passed from the daguerreotype to film to digital media. In each realm, the impulse to create and to discover, made urgent by practical challenges and opportunities, secured progress. Once the first adopters experimented with the new methods, tweaking and perfecting them along the way, it wasn't long before the old tools were left behind. In so many cases, we're all left wondering how we got by without the new technologies and why it took so long to get to them.

Innovators have dramatically changed how we lived our lives in the previous centuries. In Chicago, where human and animal waste coated streets and incubated diseases in the nineteenth century, innovators developed sewers, requiring men to use jackscrews to lift buildings and even entire city blocks to install diversion channels for the waste. Years later, innovators cordoned off the waste even more effectively with septic systems and flushable toilets. Innovators found ways to deliver light through candles, kerosene lamps, then photovoltaics, and now LEDs. They also developed air conditioners, which not only made home and business life more comfortable, but also "enabled dramatic changes in human settlement," with millions of people moving from the Rust Belt to the Sun Belt. You can make do without sewers and toilets, reservoirs or refrigerators, lights and air conditioners, but all of them make life more bearable, comfortable, and livable.

Of course, innovation follows other human impulses as well, yielding assault weapons, confinement systems, explosive-tipped harpoons, trawlers, and stereotaxic devices in animal research laboratories that have all allowed us to exploit animals more efficiently and ruthlessly than ever. Technology itself is morally neutral. It's always been a question of human motive. A GPS tracking tool can be used to chase down whales for harpooning or to study these majestic creatures for conservation purposes. An automatic weapon can be wielded just as readily by a warden or a poacher. Selective breeding or gene sequencing can be used to treat genetic health risks in dogs or to accentuate characteristics in purebreds that saddle them with shortened lives and physical abnormalities, such as elongated backs or pushed-in faces or noses. A tractor trailer can be used to transport distressed and spent dairy cows to a slaughterhouse or to transport dogs rescued from a puppy mill to safety. Technology can be the worst thing in the world for animals or the best.

When I was a kid, Kodak was synonymous with cameras and photography, just like Xerox was with photocopying. The Rochester, New York–based company was dominant in its sector, grossing $30 billion in revenues selling cameras and film and controlling the printing of the photographs. In its advertising, the company reminded people that they could capture a "Kodak moment," logging their family's history through photography. While professionals taking portraits in studios dominated photography in prior decades, Kodak opened up photography to the masses and brought it from studios into family life. As a result of Kodak's efforts, women went on to become the most lucrative market segment for photography. That loyal contingent of female customers funded Kodak's business success: they took more pictures than everyone else, printed them, shared them with friends, saved them in albums, and displayed them in the living room.

But with the Internet, there was a coincident rise in digital technology. Kodak did not adapt as this technology gained in quality and acceptance, and digital cameras eliminated the need for developing film. While Kodak clung to a business model that had served it well for decades, competitors invested in digital technologies, and by 2012 Kodak was bankrupt. Of the roughly 200 buildings clustered on Kodak's 1,300-acre campus, 80 have been demolished and 59 others sold off. Kodak has been trying to come back from bankruptcy, and now has a market capitalization of about $800 million. Meanwhile, newcomer GoPro, a maker of digital cameras for extreme sports, is worth about six times as much. Even the biggest companies must adapt, because if there's any watchword to our human experience, it's change.

We are in the midst—much closer to the beginning than to the end, I believe—of an epic political, cultural, and economic realignment in the treatment of animals. Doing something about a moral problem requires first identifying it and then intentionally breaking old habits and conventions. Confronting terrible injustices—from slavery and child labor to segregation to gender discrimination—was a painful and necessary part of our American tradition. With the availability of information on the Web, and the transparency it brings, it's harder now to sidestep these questions. Just as people are shaken from their comfort zone when they realize that a T-shirt from a well-known clothing company was stitched together in some hellish, overcrowded factory in Bangladesh, more of us are connecting our choices and purchasing practices when it comes to matters of animal cruelty. Exposing abuses goes a long way in prompting sellers to get their supply chains in order, even as it better informs their customer base. Smart businesses want to get ahead of controversy and avoid protests, boycotts, and social media campaigns that target them.

We are seeing transformations in all sectors of the animal-use economy, and we will see many more. Once resolve has set in to change for the better, it's easier than ever to make it happen. It's not a matter of sacrifice—just conscious, better choices. Enlarging our vocabulary from "whaling" to "whale watching" is just one example of an emerging shift in the manner we humans regard animals—whether in the wild, on the farm, in laboratories, or in our homes.

The society we have now is different from what it was 25 or 100 or 200 years ago, with revolutionary advances in commerce, banking, currency, energy exploration, global transport, information technology, and computing. How can we not have a commensurate revolution in our treatment of animals? How can we tolerate the misery that comes from whaling, factory farming, trapping, and cruel industries of every kind once we recognize a more vibrant economic path forward, producing jobs that are better to hold, goods that are better to have, and a society in the end that is better to live in, without unlighted places where cruel things are permitted?

A transformed commercial economy and widely embraced behavioral changes are a large part of the solution, but they won't be enough. Because of the asymmetry in power we hold over animals, even a small number of people can do terrible damage, and that's why we need laws to prevent the outliers from doing harm. In South Africa in 2014, poachers killed more than 1,200 rhinos, out of a total of 15,000 surviving there, and throughout Africa, marauding bands of poachers slaughtered 100,000 African elephants from 2010 to 2012, all to supply the global demand for "white gold." We need laws to stop the killers and the killing, and we need laws in consuming countries that forbid the sale of these products—even a few tens of thousands of American, Chinese, or Vietnamese consumers seeking ivory or rhino horn trinkets provide poachers with the financial incentive to threaten the existence of the species.

Only a small subset of hunters patronize captive hunting facilities, but even 10 percent of the hunters interested in this type of guaranteed killing have grown the industry to the point where thousands of facilities operate in the United States and South Africa. In the last decades, the HSUS has helped to pass more than one thousand state laws and dozens of federal laws, and is working to see that these laws are enforced, in order to deal with the outliers and also to establish meaningful standards in society. It's now a felony in every state to commit malicious acts of cruelty, but we need laws in other areas to address many long-standing forms of exploitation. These laws shield animals from abuse, but they also normalize the notion that cruelty to any kind of animal is unacceptable.

Animals think and feel and carry with them the spark of life. We can see as much. And ethologists and cognitive scientists can prove it. That's as much a reality as Galileo's once heretical idea that the Earth revolves around the sun rather than the other way around. Once awakened and informed about animals, there's nothing to stop us from demonstrating our particular creative genius to find more humane ways to do business. Steven Pinker, in *The Better Angels of Our Nature*, provides a long list of areas—from literacy to longevity to violent crime to democratic governance—where society is doing better and becoming better. When Pinker was asked about the few areas in which he thought we are falling short, he cited prison conditions, notably solitary confinement, and animal cruelty, particularly factory farming. Since 1900, American life expectancy has progressed from forty-seven years to about eighty. I wrote this book during my forty-ninth year, and if I had done so in 1900, I would have been living on borrowed time. But now I'm considered just middle-aged, you can read my book in digital or print form, and I can do a satellite media tour and appear on a dozen or more local stations on the same day. That's the pace of change in the modern world. Apply it to the old and exploitative industries that cause most animal suffering, and many of them can be rid from the earth even in our own lifetimes.

What will the innovations of the future be? We can use our imaginations in the search for answers. We can look to yesterday for clues, for the humane economy has a history as well as a future. The past tells us that overcoming cruelty is an ancient and noble human aspiration. The present reminds us that we now have unequaled power to achieve that goal.

This much is sure: as the humane economy asserts its own power, its own logic, and its essential decency, an older order is passing away, and the near-universal reaction to each new step will be "good riddance." By every measure, life will be better when human satisfaction and need are no longer built upon the foundation of animal cruelty. Indefensible practices will no longer need defending; unnecessary evils will no longer need excuses. In their place, in market after market, we'll see the products of human creativity inspired by human compassion, a combination that can solve any problem and overcome any wrong.

Piece by piece, purchase by purchase, choice by choice, humanity will cast off things not worth missing, and replace them with things that do us credit and show us at our best. The things we hold dear—prosperity, security, family, relationships, our health, and more—will remain. We'll just shed the cruelty, and wonder later on why it took us so long, as it's been the case with so many other moral problems of humanity's own making. In the fullness of time, we'll become more alert to animals, more appreciative of their goodness and their beauty, and more grateful, as we should be, for how they fill the world with sounds, colors, and sights that enrich every one of us in more ways than we know.

Postscript

Just weeks before this book went to press, I received a call from California's former Republican congressman John Campbell, with whom I'd worked closely for years on animal issues. Though he had no idea I had just finished a book in which I spent several thousand words about the controversy over SeaWorld, he suggested I talk to the company's new CEO, Joel Manby. Campbell and Manby went back a long time, connecting originally during their prior careers in the automobile business; more important, Congressman Campbell told me he knew Manby to be a strong leader—one who might just be the person to change SeaWorld's archaic business model.

Playing the role of a diplomat bringing enemies to the negotiating table, Campbell asked if I'd agree to a sit-down with Manby. Naturally I was conflicted about meeting with the leader of a company that had long earned harsh criticism from animal advocates, and it felt especially awkward since I had just written an intentionally hard-hitting chapter about SeaWorld for this book in which I detailed their long and unfortunate animal-welfare record. And yet I have always believed in engaging with our adversaries to find a path forward that will produce the best outcomes for animals. So I agreed to the face-to-face meeting; with Campbell vouching for me, Manby took his own leap of faith and agreed to do the same.

As soon as Manby and I began our talks, I felt confident Congressman Campbell was right about him. I recognized Manby as a leader with the ability to guide SeaWorld in the years ahead and make a sharp turn toward the humane economy. Over the next few weeks, we entered into discussions, and then in spring of 2016, SeaWorld announced a set of reforms, with HSUS lauding the moves while still reminding the public that we'd continue to have some differences with the company.

With the support of his board, Manby pledged that SeaWorld would stop breeding orcas, making the current group of twenty-nine whales the last generation to live and work there. The agreement also stipulates that no new SeaWorld parks anywhere in the world will have orcas. What's more, SeaWorld has pledged to invest millions of dollars in rescuing and rehabilitating marine mammals, including using some of the nonreleasable animals for needs at the exhibits that would still have live animals. He also agreed to work with HSUS to campaign to end the commercial killing of whales and seals around the world and to end shark finning—turning SeaWorld's parks into education and advocacy centers for the oceans' creatures. With more than 20 million people passing through the gates of its facilities every year, I knew that could make a major difference in our global campaigns to protect marine mammals.

Manby and his team proved that with the right people stepping up and recognizing the need for reform, even businesses long at odds with animal-protection groups can find a way forward and step into the humane economy. Not only would SeaWorld's business model change for the better and improve the lives of orcas and other captive marine mammals, but we'd also won an important ally in our fight for millions of wild marine creatures who need all of us more than ever.

Indeed, this sort of movement is precisely the way the new humane economy is built—with consumers gently and sometimes urgently pushing companies to do more for animals. When the companies take tangible steps in the right direction, as difficult as it may be for them at the start, they liberate not only the animals, but also their economic potential. They shed the protest and the acrimony and the lawsuits and align themselves with the mass of consumers who want to know—no matter whether it's live entertainment, food and agriculture, or laboratory science or wildlife management—that there is a better, more sustainable, and more humane way. When moral reasoning and economic opportunity are aligned, the conditions are ripe for a company to thrive alongside our extraordinary animal cousins.

Ten Things You Can Do to Contribute to the Humane Economy

1. **Vote with your dollars**
   Every time you enter the marketplace, you vote for or against cruelty with your dollars. Choose cosmetics and cleaning products not tested on animals. Adopt dogs or cats from rescues or shelters, don't buy animals from pet stores, and never keep a wild animal for a pet. Avoid fur, exotic leathers, and the other cruel products of the commercial wildlife trade. Search our website, www.humanesociety.org, for buying guides focused on these issues.

2. **Eat lower on the food chain**
   Make your food choices a referendum on factory farming. Nourish yourself by eating more plant-based foods, which are better for you, the planet, and animals. Reduce your consumption of meat. If you eat animal products, choose those with the labeling certifications "Global Animal Partnership," "Animal Welfare Approved," or "Certified Humane."

3. **Join and support the HSUS and other groups that help animals**
   There is no substitute for organized, collective action for animals. Animal organizations rely on your support to advocate for progress in the realms of public and corporate policy and to foster

greater understanding of humane issues. They mobilize their members and supporters to build awareness, to advance their campaigns, to provide direct care for animals, and to apply pressure on lawmakers, corporations, and other decision-makers.

4. Influence lawmakers

Public officials act when they hear people clamoring for change. Contact lawmakers at the local, state, and federal level about pending animal welfare legislation. Ask them to adopt animal-welfare platforms. Enacting laws against malicious animal cruelty, animal fighting, killing marine mammals, stopping extreme confinement of farm animals, and banning trophy hunting of carnivores has been transformative for animals. Sign up for email alerts at www.humanesociety.org.

5. Urge enforcement of our laws

Laws are of limited value if not enforced. Encourage law enforcement agencies to crack down on cruelty, and press mayors, governors, and other officials to enforce the law. You can also write to executive agencies, such as state and federal agriculture departments and wildlife agencies. For instance, in 2015, the California Fish and Game Commission banned the commercial trapping of bobcats after tens of thousands of citizens wrote and urged that action.

6. Minimize and eliminate food waste.

Americans throw out 40 percent of their food (including 22 percent of animal products), and there are enormous costs, in the form of greenhouse gas emissions and energy inputs, in producing food. We could spare more than a billion animals in the United States alone every year just by eliminating food waste. Ask restaurants to provide appropriate portions and take what food you don't eat home so you can eat it later. Split a plate if you and your dining companion know that will be enough for you at that meal.

7. Share information about animal protection

The greatest antidote to cruelty is an informed citizenry. There's so much good information on the Web about animal issues—find it, access it, and share it with friends and family members on Facebook, Twitter, and other social media. When you give a gift, consider a book, film, or some other item that encourages them to learn more about animal protection.

8. Remind corporate leaders to do the right thing

If you are an investor in a company, write a letter to the CEO about establishing animal-welfare policies. Urge the leadership at the company to be conscious about animal-welfare issues and urge it to adopt best practices for its industry. If you are a member of a pension fund, urge its managers not to invest in companies engaged in factory farming or meat processing, and if they do, ask that they at least insist upon meaningful animal-welfare policies.

9. Travel like an eco-tourist

Pursue eco-friendly destinations when you take a vacation. This brand of tourism drives the global economy and benefits nations whose commitment to the needs of wildlife requires habitat preservation and management, law enforcement and ranger work, and other necessary services to protect wild animals and their visitors.

10. Live as if all life matters

Adopt pets or foster them. Pick up plastic waste on beaches and in parks. Minimize your own waste, and dispose of it carefully, since that waste not only eats up space and wildlife habitat but can kill animals. Buy a fuel-efficient vehicle, and bike and walk more.

Acknowledgments

We wouldn't see the remarkable changes for animals sweeping through our society and economy without passionate, caring people driving this revolution in word and deed. As the business leaders, politicians, and so many other key decision-makers involved would attest, their work stands on the shoulders of millions of conscious consumers and voters. I thank every one of you who raises your voice for animals; future progress depends on your continued diligence and impatience with the status quo.

I have an immeasurable debt to one person—Audrey Steele Burnand, an advocate and philanthropist I met during the Proposition 2 campaign in California in 2008 to end extreme confinement of farm animals and who was instrumental in its success. Our common devotion to securing that reform has developed into a very special friendship and partnership. Audrey has enabled so many of the extraordinary changes that I've written about in this book, influencing the course of history in profound ways. And she's done all of it after she reached her eighty-seventh birthday. She was always prepared to take on the tough fights, and she understood the necessity of finding the resources to wage those battles. She intuitively knew that we cannot rescue our way out of the problems for animals, but that we need fundamental change and must prevent cruelty before it occurs. I dedicate this book to her—a cherished friend and a compatriot in this battle to find a safe place for every beating heart.

I am grateful to all of my colleagues at the HSUS, the greatest assembly of talent ever pulled together on one team to fight for animals, including our COO and my comrade Michael Markarian. When it came to writing this book, I am so appreciative of the efforts of our farm animal team, especially Paul Shapiro, Josh Balk, and Matt Prescott; our animal research team of Andrew Rowan, Katie Conlee, Troy Seidle, and Kate Willett, with a great assist from Sara Amundson; our companion animal experts Betsy McFarland and Carrie Allan; and our wildlife advocates Stephanie Boyles Griffin, Teresa Telecky, Iris Ho, Kitty Block, Josh Irving, Nicole Paquette, John Hadidian, John Griffin, Lisa Wathne, and Debbie Leahy. I am also grateful to Michelle Cho, Jon Lovvorn, and John Cleveland for their invaluable assistance. Erich Yahner provided tremendous support in the effort to pull the notes for the book together. Rachel Querry, whom I've worked with for more than twenty years at HSUS, has been a constructive force on both of my books and a fantastic promoter of the works. Our general counsel Roger Kindler gave me first-rate legal advice, and he's as smart as they come. All of these folks read portions of the manuscript, gave me their insights, and did their best to keep it clean of any errors. Any mistakes are mine alone.

I'm also grateful to the board of directors of the HSUS, including its chairman Rick Bernthal. Immensely dedicated to the cause, he possesses an uncommon social intelligence and common sense, and his instincts on the book were helpful at every stage. I feel a special gratitude to Rick, our vice chairs Jason Weiss and Jennifer Leaning, our prior board chairs David Wiebers and Anita Coupe, and to the rest of the board for giving me the extra runway to get this project up in the air and then to land it safely, knowing that I have substantial other duties to execute for the HSUS. Every one of our board members knows that a serious-minded group driving transformational outcomes needs to show thought leadership and be involved in writing and publishing, and I am just one of the members of the HSUS family who has benefited from that understanding.

I am also so appreciative of the tireless, intelligent work of my adviser Crystal Moreland. She is everything you want in a right-hand person—loyal, responsible, and intuitive. She's got a bright future in the animal-protection movement, and she's a joy to work with.

It was my good fortune to interact directly with so many of the people you've read about in this book. I am grateful to Darren Aronofsky, Rick Jaffa and Amanda Silver, and Gabriela Cowperthwaite, who spoke about their creative work within the movie business and its intersection with animal protection. I am indebted to Stella Trueblood, Jerome Fox, and Kayla Grams for spending the day with me in an extraordinary landscape in far northwestern Colorado, and I so admire their work in helping wild horses. I wish to thank Chris Schindler and the other members of our animal rescue team on that raid of the dogfighting operation in southern Alabama; it's heartbreaking to see animals in terrible circumstances but it still buoys me to know we pulled them out of that mess. I cannot thank wildlife scientists Rolf Peterson and John Vucetich enough for spending several days with me at Isle Royale National Park, sleeping in the wilderness and one day even hiking for nearly twenty miles. Rolf is upwards of sixty-five, but you'd never guess it from the vigor he showed on the trail. Thanks, too, to researchers Don Ingbar and Chris Austin and their colleagues, who allowed me to witness and learn more about their innovative work in medical science. Ron Kagan gave me critical input all along the way.

Several years ago, my colleague Sarah Barnett invited me and my wife, Lisa, to meet a beautiful bunch of homeless dogs at a PetSmart in Virginia. That encounter not only provided an inspiration for one section of the book, but connected me with my little Lily, the beagle mix I cannot imagine living without. Christina Morgan was wonderful in showing me her pet adoption store and talking to me about the challenges facing animals in her community. Jerry Crawford spent many days with me in the United States and in Europe visiting egg farms; we learned so much from each other, and out of it has come an enduring friendship. I was heartened by my visit to Marcus Rust's farm in Indiana, and thank him for opening up his doors, especially after some prior conflict between our corporations. Josh Tetrick and Ethan Brown, entrepreneurs and animal advocates, showed me what they do each day at work, and Andras Forgacs was generous in sharing his ideas about the future of food. My thanks to Simon Belcher, too, for talking to me about the one day in his life that he'd most like to forget.

In my work, I have interacted with many dedicated politicians and officials. It was great fun to spend time with Senator Gary Peters at Isle Royale, especially so because we have worked together to protect wolves and many other creatures. Senator David Vitter spent a full day with me touring the southern coast of Louisiana after the Deepwater Horizon spill and has fought bravely in the US Senate for so many animal-protection policies—I am in his debt. Senator Cory Booker is one of a kind, and he's bringing his leadership, energy, and smarts to the cause of animal protection, which will never be the same with him on the team. Representative Earl Blumenauer understands the intersection of animal protection and other social concerns, and he's emerged as the national thought leader on animals and politics; I've learned so much from him.

On the writing and editing side, I had a tremendous and talented support group. I am lucky to have a first-rate editor at William Morrow in Peter Hubbard. Peter edited *The Bond*, and I felt very lucky indeed to get this second run with him. He has an uncanny macro-level view of how a book should flow, and when it's time to move on from one topic to the next. My agent Gail Ross, who first connected me with Peter, does not take a passive role once the book contract is done. She helped to make the humane economy concept more granular, making this work, I hope, a more appealing one to a mainstream audience.

Once I finished a draft chapter, I first handed it off to Lewis Bollard, who worked side by side with me at the HSUS. Lewis, who graduated from Harvard College and then Yale Law School, is doing extraordinary work for the cause and has immense potential as a future leader of the animal-protection movement. He is wise beyond his years, with a lightning-fast mind and off-the-charts skills. He was tireless in reviewing my work, offering line edits and providing encouragement for so much of the on-the-ground storytelling that appears in the book.

I am also deeply grateful to John McConnell and Matthew Scully, former White House speechwriters, for reviewing the manuscript. John and Matthew live in the world of ideas, economics, and policy, and as writers, they have an unmatched sense for how to get those concerns across to the reader. Matthew and I have been the closest of friends for nearly twenty years, and his immense talent, which was so on display in his own book *Dominion,* was essential to my thinking about both *The Bond* and *The Humane Economy*.

John Balzar, a writer with a long, distinguished career in journalism at the *Los Angeles Times* before coming to work with me at the HSUS, also reviewed every chapter. He possesses both a big-picture view of the world and an extraordinary ability to communicate about it in a crisp and powerful way. John especially helped me think about framing the entire work, and his presence is felt throughout the book.

My deepest thanks, too, to my longtime colleague and friend Bernard Unti, a professional historian and partner in this cause for thirty years. He knows the history of our cause as well as anyone, and it's invaluable to have an asset like that at the ready. Bernie was thoroughly engaged throughout this process and improved every aspect of this work with his insight and multidisciplinary talents.

Not a day goes by when I do not have occasion to think of the role my family members have played in my life. It hardly seems enough to say that I am grateful to my parents, Pat and Richard Pacelle, for their support, their compassion, and attention to my career, just as I am grateful to them for their personal examples of strength, principle, and love. My brother, Richard, has always been my greatest inspiration, and I am so fortunate to have two very caring and wonderfully successful sisters, Kim and Wendy. As a family, we have always celebrated one another's successes, and we have always stood together.

Finally, this book would not have come together as it did without my wife and her presence in my life. Lisa Fletcher is a top-notch journalist and one of the most talented and well-rounded people I know. She can do just about anything, and does all of it with humility and vigor. She married a man whose job is one with unremitting pressure and demands, and she's my biggest supporter, and I'm so grateful for her understanding and love. She's got a very discerning eye and helped me at so many stages of this process, right up to the end. I love her, and am so lucky to have her in my life.

Thanks again to each of you for taking the time to explore the ideas I have presented in this work. While I may not know or meet you, I think of you as co-creators of the humane economy and a safer world for everyone.

Notes

Introduction

xi the eminent economist Joseph Schumpeter: Schumpeter, Joseph, "The Process of Creative Destruction," *Capitalism, Sociology, and Democracy*, New York: Harper Colophon Edition, 1976, p. 84.

ONE: Pets and GDP (the Gross Domesticated Product)

6 Virginia is one of the East's major mill states: "Puppy Mills Then and Now: A Decade of Progress." The Humane Society of the United States, 2009, p. 5. http://www.humanesociety.org/assets/pdfs/pets/puppy_mills/report-puppy-mills-then-now.pdf

10 Total sales within the US pet sector: Associated Press, "Americans Spent a Record $56 Billion on Pets Last Year," *CBS Money Watch*, March 13, 2014, http://www.cbsnews.com/news/americans-spent-a-record-56-billion-on-pets-last-year/.

11 BC Partners . . . acquired PetSmart for $8.7 billion: de la Merced, Michael, "PetSmart Accepts $8.7 Billion Buyout," *The New York Times*, December 14, 2014, http://dealbook.nytimes.com/2014/12/14/petsmart-to-sell-itself-to-investor-group-for-8–7-billion/?_r=0.

11 PetSmart and Petco have helped to transfer: Petco Foundation, "Online Adoption Search," n.d., http://www.petco.com/petco_Page_PC_petfinderadoptions.aspx; PetSmart, "Saving Lives through Pet Adoption," n.d., http://pets.petsmart.com/adoptions/.

13 Today, about 12,000 shelters and rescues upload photos and information: Petfinder, "Animal Shelters & Rescues," n.d., https://www.petfinder.com/animal-shelters-and-rescues/.

13 more than 250,000 animals pictured on the site: Petfinder, "About Petfinder," n.d., https://www.petfinder.com/about/.

13 Petfinder.com has made possible more than twenty-two million pet adoptions: Chudgar, Sonya, "Nestle Purina to Buy Petfinder," *Advertising Age*, June 10, 2013, http://adage.com/article/news/nestle-purina-buy-petfinder/241995/.

14 David Duffield . . . has committed more than a billion dollars to his foundation:

"David Duffield," *Inside Philanthropy*, n.d., http://www.insidephilanthropy.com/guide-to-individual-donors/david-duffield.html.

15 there are 3,500 brick-and-mortar shelters: Humane Society of the United States, "US shelter and adoption estimates for 2012–13," Pets by the Numbers, January 30, 2014, http://www.humanesociety.org/issues/pet_overpopulation/facts/pet_ownership_statistics.html.

15 collectively spending $2.5 billion a year on solving the problem of animal homelessness: Personal communication, Andrew N. Rowan, November 2015.

15 these shelters employ a workforce of thirty-five thousand people: Personal communication, Andrew N. Rowan, November 2015.

15 there are another 13,000 plus rescue groups: Personal communication, Andrew N. Rowan, November 2015.

16 all receive a tenth of available resources, even though they represent more than 99 percent of the animals at risk: Personal communication, Andrew N. Rowan, November 2015.

17 some facilities are known to conduct as many as twenty-three thousand cat and dog sterilizations every year: Peters, Sharon, "Humane Alliance Leads Way in Low-Cost Neutering and Spaying," *USA TODAY*, April 27, 2010, http://usatoday30.usatoday.com/life/lifestyle/pets/2010–04–28-humanealliance28_ST_N.htm.

17 The Alliance has now helped 137 start-up spay-and-neuter clinics: Niedziela, Ken, "Humane Alliance Spay/Neuter School Expands," *Veterinary Practice News*, July 30, 2014, http://www.veterinarypracticenews.com/Humane-Alliance-Spay-Neuter-School-Expands/.

17 83 percent of owned dogs and 91 percent of owned cats are now spayed or neutered in the United States: Quenqua, Douglas, "New Strides in Spaying and Neutering," *The New York Times*: Well Pets Blog, December 2, 2013, http://well.blogs.nytimes.com/2013/12/02/new-strides-in-spaying-and-neutering/?_r=0.

18 Dr. Gary Michelson . . . is providing more than $10 million a year: Darcé, Keith, "Sign On San Diego: Drugmaker Works to Sterilize Pets Without Surgery," Found Animals, n.d., http://www.foundanimals.org/about-us/news/drugmaker-works-sterilize-pets-without-surgery.

18 Researchers have already developed contraceptive vaccines for more than seventy wildlife species: Kirkpatrick, J.F., R.O. Lyda, & K.M. Frank, "Contraceptive Vaccines for Wildlife: A Review," *American Journal of Reproductive Immunology*, July 2011, 66, 40–50.

18 Zeuterin . . . is currently the only nonsurgical sterilant with U.S. Food & Drug Administration approval: Graves, Kandace, "The Future of Non-Surgical Neutering for Dogs," *Gambit*, March 25, 2014, http://www.bestofneworleans.com/gambit/a-whole-new-ballgame/Content?oid=2409983.

18 Rose Wilson . . . told *The Wall Street Journal* that the drug is "simple, it's inexpensive, and it's painless . . .": Beck, Melinda, "Too Many Dogs: A Simple Solution," *The Wall Street Journal*, November 28, 2014, http://www.wsj.com/articles/too-many-dogs-a-simple-solution-for-sterilization-1417187544.

19 Since the mid-1970s, the number of dogs and cats euthanized has fallen: Mach, Andrew, "Behind the Big Drop in Euthanasia for America's Dogs and Cats," *The Christian Science Monitor*, February 10, 2012, http://www.csmonitor.com/USA/

Society/2012/0210/Behind-the-big-drop-in-euthanasia-for-America-s-dogs-and-cats.

24 Only 1 percent of American pets have healthcare plans: Brady, Diane, and Christopher Palmeri, "The Pet Economy," *Bloomberg Business*, August 5, 2007, http://www.bloomberg.com/bw/stories/2007–08–05/the-pet-economy; "History of Pet Insurance," FiloFETCH, November 10, 2005, http://filofetch.com/history-of-pet-insurance/.

25 "She was under three pounds . . .": Avellino, Kelly, "Bailey's Law Aims to Curb VA Pet Shops from Dealing with Puppy Mills," NBC 12, January 23, 2014, http://www.nbc12.com/story/24534508/baileys-law-aims-to-curb-va-pet-shops-from-dealing-with-puppy-mills.

25 Her story inspired a namesake law in Virginia: Avellino, Kelly, "Bailey's Law Aims to Curb VA Pet Shops from Dealing with Puppy Mills," NBC 12, January 23, 2014, http://www.nbc12.com/story/24534508/baileys-law-aims-to-curb-va-pet-shops-from-dealing-with-puppy-mills.

26 Petland . . . has been heavily criticized by the humane community: Munguia, Hayley, "New Crackdowns On Breeding Won't Mean Your Pet Isn't From A Puppy Mill," FiveThirtyEight, October 24, 2014, http://fivethirtyeight.com/datalab/new-crackdowns-on-breeding-wont-mean-your-pet-isnt-from-a-puppy-mill/.

26 more than thirty of its one hundred or so stores have shut down: Petland, "Search U.S. Petland Stores," Petland Stores, n.d., http://www.petland.com/stores/map/index.html.

26 In 2014, Pets Plus Natural . . . adopted a "puppy friendly" policy to replace puppy sales: Worden, Amy, "Philly Area Pet Stores Go 'Puppy Mill Free,' Some Fear Rise in Backyard Breeding," philly.com, November, 24, 2014, http://www.philly.com/philly/blogs/pets/Philly-area-pet-stores-go-puppy-mill-free.html#UcB6Ck1c2AvzPGoM.99.

27 Chicago, Los Angeles, Miami, New York, Phoenix, and more than 100 other big cities: Best Friends Animal Society, "Jurisdictions with Retail Pet Sale Bans," Resources, n.d., http://bestfriends.org/resources/jurisdictions-retail-pet-sale-bans.; Good, Kate, "Paws Up! NYC Enacts Law to Ban the Sale of Puppy Mill Dogs," One Green Planet, n.d., http://www.onegreenplanet.org/news/nyc-enacts-law-to-ban-the-sale-of-puppy-mill-dogs/#.

29 All told . . . animal welfare groups rescued 367 dogs: Rawls, Phillip, "Dog Fighting Ring Busted: 367 Pit Bulls Rescued," *The Christian Science Monitor*, August 27, 2013, http://www.csmonitor.com/USA/Latest-News-Wires/2013/0827/Dog-fighting-ring-busted-367-pit-bulls-rescued.

29 It is thought to be the second-largest dogfighting bust ever in the United States: Federal Bureau of Investigation, "Strong Sentences Handed Down by Alabama Court in Historic Dog Fighting Case," November 12, 2014, https://www.fbi.gov/mobile/press-releases/2014/strong-sentences-handed-down-by-alabama-court-in-historic-dog-fighting-case.

30 we joined law enforcement teams and other animal groups in seizing more than five hundred dogs: Katz, Neil, "Inside America's Biggest Dog-Fighting Bust," *CBS News*, July 9, 2009, http://www.cbsnews.com/news/inside-americas-biggest-dog-fighting-bust/.

30 we'd helped to rescue more than four hundred dogs: Associated Press, "Judge Hopes Dog Fighting Sentences Send Warning," Fox 10 TV, n.d., http://www.fox10tv.com/story/27814154/judge-hopes-dog-fighting-sentences-sends-warning.

30 Ten suspects . . . were arrested and indicted on felony dogfighting charges: WTVY, "Update: Names Released in Multi-State Dog Fighting Case," August 26, 2013, http://www.wtvy.com/news/alabama/headlines/367-Dogs-Rescued-in-Multi-State-Dog-Fighting-Case-221175641.html.

30 Federal and local officials also seized: WTVY, "Update: Names Released in Multi-State Dog Fighting Case," August 26, 2013, http://www.wtvy.com/news/alabama/headlines/367-Dogs-Rescued-in-Multi-State-Dog-Fighting-Case-221175641.html.

30 "The lowest places in hell would be reserved for those who would commit cruelty to animals": Sayers, Devon, and Joe Sterling, "Feds: Busted Dogfighters Earn Spot in 'lowest places in hell,' " CNN, August 27, 2013, http://edition.cnn.com/2013/08/26/justice/alabama-dog-fight-ring/.

31 It's still legal in more than 120 countries: Personal communication with HSI vice president Kitty Block.

31 animal welfare groups had to hold almost four hundred dogs for evidentiary purposes: Associated Press, "Judge Hopes Dog Fighting Sentences Send Warning," Fox 10 TV, n.d., http://www.fox10tv.com/story/27814154/judge-hopes-dog-fighting-sentences-sends-warning.

32 The puppy mill operators even sued the USDA: Texas Humane Legislation Network, "Dogs Victorious in Legal Battle Over New Puppy Mill Law," Animal Protection Issues, July 23, 2013, http://www.thln.com/?pageID=12F33A77–3048-C277–11BAACD1FAC24A23&theme=Puppy%20Mills.

32 Toward the end of 2014, a federal judge meted out the most severe penalties: Burylo, Rebecca, "8 Connected to Dog-Fighting Ring Get Lengthy Prison Terms," Montgomery Advertiser, November 13, 2014, http://www.montgomeryadvertiser.com/story/news/crime/2014/11/13/dog-fighting-ring-given-long-prison-sentences/18976433/.

32 Donnie Anderson . . . received an eight-year sentence: Edgemon, Erin, "Auburn Man Sentenced to 8 Years in Prison for Role in High-Stakes Dog Fights," AL.com, November 7, 2014, http://www.al.com/news/montgomery/index.ssf/2014/11/auburn_man_sentenced_to_8_year.html.

32 Several others . . . received stern sentences: Edgemon, Erin, "Auburn Man Sentenced to 8 Years in Prison for Role in High-Stakes Dog Fights," AL.com, November 7, 2014, http://www.al.com/news/montgomery/index.ssf/2014/11/auburn_man_sentenced_to_8_year.html.

TWO: Big Ag Gets Its Hen House in Order

34 With assets estimated at $25 billion: *Forbes*, #31 Carl Icahn, The World's Billionaires, n.d., http://www.forbes.com/billionaires/list/.

34 The company buys about 1 percent of US pork: Strom, Stephanie, "McDonald's Set to Phase Out Suppliers' Use of Sow Crates," *The New York Times*, February 13, 2012, http://www.nytimes.com/2012/02/14/business/mcdonalds-vows-to-help-end-use-of-sow-crates.html?_r=0.

34 but about 15 percent of all US pork bellies, which means it takes a piece of one in seven of the nation's pigs: Personal communications, Bob Langert, McDonald's vice president, January 2012.

35 "Forget the pig is an animal . . . Treat him just like a machine in a factory": Robbins, John, "The Most Unjustly Maligned of All Animals," *Diet for a New America* (25th Anniversary Edition), 2012, 63.

35 in 1980, there were nearly 700,000 pig farmers, and today there are fewer than 60,000: USDA, Crop Reporting Board, Economics and Statistics Service, Washington, D.C. December 23, 1980. Hog and Pig Farming, USDA, Census of Agriculture Highlights, Washington, D.C., June 2014. More details about the dramatic drop in the number of farmers can be found in Leonard, Christopher. *The Meat Racket: The Secret Takeover of America's Food Business* (Simon & Schuster, New York, 2014).

38 "McDonald's believes gestation stalls are not a sustainable production system for the future . . . There are alternatives that we think are better for the welfare of sows . . .": Strom, Stephanie, "McDonald's Set to Phase Out Suppliers' Use of Sow Crates," *The New York Times*, February 13, 2012, http://www.nytimes.com/2012/02/14/business/McDonalds-vows-to-help-end-use-of-sow-crates.html?_r=0.

38 "This is a business decision that they made . . . . We have no problem with that . . .": Smith, Aaron, "McDonald's Phasing Out Tiny Cases for Pigs," *CNN Money*, February 15, 2012, http://money.cnn.com/2012/02/14/news/companies/McDonalds_pigs/.

39 "So our animals can't turn around for the two-and-a-half years that they are in stalls producing piglets . . . I don't know who asked the sow": Tepper, Fabien, "The New Ethics of Eating," *The Christian Science Monitor*, December 7, 2014, http://www.csmonitor.com/USA/Society/2014/1207/The-new-ethics-of-eating.

41 American per capita consumption of veal dropped from 8.6 pounds to just 0.3 pounds: Calculated by ERS/USDA based on data from various sources (see http://www.ers.usda.gov/data-products/food-availability-(per-capita)-data-system/food-availability-documentation.aspx).

41 The veal industry trade group . . . finally pledged in 2007 to nix the stalls: American Veal Association, "Ever Wonder What Veal is or Where it Comes From?" *Veal FAQ*, n.d., http://www.americanveal.com/for-consumers/veal-frequently-asked-questions/.

41 Walmart . . . accounting for 25 percent of all grocery sales in the United States: Mitchell, Stacy, "Will Wal-Mart Replace the Supermarket?" *Salon*, March 28, 2013, http://www.salon.com/2013/03/28/will_wal_mart_replace_the_supermarket_partner/.

42 "There is a growing public interest in how food is produced . . .": Walmart, "Animal Welfare—Swine Assurance Position," *Walmart Policies and Guidelines*, n.d., http://corporate.walmart.com/policies.

42 The company said it was "committed to continuous improvement in the welfare of farm animals in our supply chain": Walmart, "Animal Welfare—Swine Assurance Position," *Walmart Policies and Guidelines*, n.d., http://corporate.walmart.com/policies.

42 offering the "Five Freedoms" of animal welfare as an ethical framework: Walmart, "Animal Welfare—Swine Assurance Position," *Walmart Policies and Guidelines,* n.d., http://corporate.walmart.com/policies.

43 "I didn't want my success or Chipotle's to be based on that": Evans, Clay, "At the Table With Steve Ells," *Coloradan Magazine*, August 19, 2011, https://www.coloradanmagazine.org/2011/08/19/at-the-table-with-steve-ells/.

44 Its revenues for the first time ever exceeded a billion dollars in the first quarter of 2015: Condensed Consolidated Statement of Income and Comprehensive Income (USD $), http://www.sec.gov/cgi-bin/viewer?action=view&cik=1058090&accession_number=0001058090-15-000012&xbrl_type=v.

44 Chipotle had done $727 million in sales in the first quarter just two years earlier: Condensed Consolidated Statement of Income and Comprehensive Income (USD$), http://www.sec.gov/cgi-bin/viewer?action=view&cik=1058090&accession_number=0001193125-13-161154&xbrl_type=v#.

44 "We would rather not serve pork at all than serve pork from animals that are raised this way . . . Replacing the supply we have lost in these ways will take some time . . .": Doering, Christopher, "Chipotle Suspends Pork Sales Over Pig Treatment," *The Des Moines Register*, January 14, 2015, http://www.desmoinesregister.com/story/money/business/2015/01/14/chipotle-suspends-pork-sales-pig-treatment/21775603/.

45 GAP was connecting millions of consumers with more than three thousand farms through more than four hundred Whole Foods stores: Global Animal Partnership, "The 5-Step® Animal Welfare Rating Program," *Improving the Lives of Farm Animals Step by Step*, n.d., http://www.globalanimalpartnership.org/.

45 Whole Foods does $16 billion in annual sales: WFM Company Financials, http://www.nasdaq.com/symbol/wfm/financials?query=income-statement.

45 There are now more than 300 million animals certified under the program: Global Animal Partnership, "The 5-Step® Animal Welfare Rating Program," *Improving the Lives of Farm Animals Step by Step*, n.d., http://www.globalanimalpartnership.org/.

46 A 2007 poll . . . found that 95 percent of Americans believe farm animals should be well-cared for . . . they also found that nearly 90 percent believe food companies that require their suppliers to treat animals well are "doing the right thing": Lusk, Jayson L., and F. Bailey Norwood. "A survey to determine public opinion about the ethics and governance of farm animal welfare." *Journal of the American Veterinary Medical Association*, Vol. 233, No. 7 (2008): 1121–1126.

46 A 2009 study . . . found that 60 percent or more of respondents in every US state with pig production would support a legal ban on gestation crates: Tonsor, G., N. Olynk, and C. Wolf. "Consumer Preferences for Animal Welfare Attributes: The Case of Gestation Crates." *Journal of Agricultural and Applied Economics*, December 2009, 713–730.

46 "I'm confident that a ballot initiative prohibiting the use of gestation stalls would pass in nearly every state in the Union": Gauldin, Cliff, "Poll Reveals Stall Ban Support," Highbeam Business (originally published by Feedstuffs), May 18, 2009, http://business.highbeam.com/409224/article-1G1-208055707/poll-reveals-stall-ban-support.

46 A ban on gestation crates in all twenty-eight European Union member states took effect in 2013: Compassion in World Farming, "The 2013 Sow Stall Ban," About the EU Sow Stall Ban, n.d., http://www.ciwf.org.uk/our-campaigns/pigs/about-the-ban/.

47 in 2014, Canada and Australia agreed to phase out the crating system: National Farm Animal Care Council, "Gestating Gilts and Sows," *Code of Practice for the Care and Handling of Pigs*, 2014, 10–11, http://www.nfacc.ca/pdfs/codes/pig_code_of_practice.pdf.

47 By 2015 . . . Smithfield Foods had converted 70 percent of its company-owned operations to gestation crate–free: Cheeseman, G., "Smithfield Foods Phases Out Sow Gestation Crates," justmeans, January 15, 2015, http://www.justmeans.com/blogs/smithfield-foods-phases-out-sow-gestation-crates.

47 In 2014, Cargill announced that it had converted all of its breeding operations from crates to group housing: Hughlett, M., "Cargill Says Conversion to More Humane Sow Farms Is Complete," *Star-Tribune*, January 29, 2015, http://www.startribune.com/cargill-says-conversion-to-more-humane-sow-farms-is-complete/290100071/.

47 Tyson Foods . . . 14.6 percent of its $37.6 billion in annual sales going through Walmart: Tyson Foods, Inc., Form 10-K (Annual Report), 2014, http://s1.q4cdn.com/900108309/files/doc_financials/2014/TSN-FY14–10-K_v001_c2s3d9.pdf.

48 "If gestation crates are not part of the lingua franca of most investors," ISS notes, "long-term risk certainly is": Brady, Diane, "The Humane Society's New Pitch: This Pork Producer Is a Bad Investment," *Bloomberg Business*, April 21, 2014, http://www.bloomberg.com/bw/articles/2014–04–21/the-humane-societys-new-pitch-this-pork-producer-is-a-bad-investment.

49 Rust and Rose Acre donated a half million dollars to the campaign to defeat Proposition 2: Ballotpedia, "Opposition," California Proposition 2, Standards for Confining Farm Animals (2008), n.d., https://ballotpedia.org/California_Proposition_2,_Standards_for_Confining_Farm_Animals_(2008)#cite_note-26.

53 In the United States, upwards of 85 percent of laying hens live in battery cage confinement systems: Baden-Mayer, Alexis, "How—and Why—to Boycott Eggs from Factory Farms," February 19, 2014, https://www.organicconsumers.org/essays/how%E2%80%94and-why%E2%80%94-boycott-eggs-factory-farms.

53 just one hundred or so companies . . . accountable for upwards of 85 percent of the 90 billion eggs sold in the country: Baden-Mayer, Alexis, "How—and Why—to Boycott Eggs from Factory Farms," February 19, 2014,https://www.organicconsumers.org/essays/how%E2%80%94and-why%E2%80%94-boycott-eggs-factory-farms.

According to USDA: "US egg operations produce over 90 billion eggs annually. Over three-fourth of egg production is for human consumption (the table-egg market). The remainder of production is for the hatching market. These eggs are hatched to provide replacement birds for the egg-laying flocks and to produce broiler chicks for growout operations. The top five egg-producing states are Iowa, Ohio, Pennsylvania, Indiana, and Texas." http://www.ers.usda.gov/topics/animal-products/poultry-eggs/background.aspx.

56 both DeCosters were sentenced to three months in jail, and fined $7 million:

Associated Press, "Egg Executives Sentenced to 3 months for salmonella outbreak," *The Des Moines Register*, April 13, 2015, http://www.desmoinesregister.com/story/money/business/2015/04/13/decoster-salmonella-outbreak-sentencing/25702501/.

62 the average American in 2014 consuming 264 eggs: American Egg Board, "The Egg Business (Graphic)." n.d., http://www.aeb.org/images/PDFs/EggBusiness515.pdf.

62 Each hen produces . . . about 260 per year: American Egg Board, "U.S. Egg Facts," The Egg Business, n.d., http://www.aeb.org/farmers-and-marketers/industry-overview.

63 Starbucks . . . agreed to source all of its eggs from cage-free farms by 2020: Huffstutter, P.J., "Starbucks to Switch to 100 Percent Cage-Free Eggs by 2020," Reuters, October 1, 2015, http://www.reuters.com/article/2015/10/01/us-starbucks-eggs-idUSKCN0RV5HM20151001#uhyxfDGbOdC2daB2.97.

63 Compass Group, Sodexo, and Aramark—also agreed to go cage free; they buy more than a billion eggs a year: Shanker, Deena, "How the Humane Society Convinced Nearly 100 Food Companies to Take Their Animals Out of Cages," *Fortune*, April 9, 2015, http://fortune.com/2015/04/09/humane-society-food-companies-negotiation/.

63 Dunkin' Donuts also weighed in and signaled a change: Shanker, Deena, "How the Humane Society Convinced Nearly 100 Food Companies to Take Their Animals Out of Cages," *Fortune*, April 9, 2015, http://fortune.com/2015/04/09/humane-society-food-companies-negotiation/.

63 the company announced it would go entirely cage free over the next decade: Strom, Stephanie, "McDonald's Plans a Shift to Eggs From Only Cage-Free Hens," *The New York Times*, September 9, 2015, http://www.nytimes.com/2015/09/10/business/McDonalds-to-use-eggs-from-only-cage-free-hens.html.

65 With the company buying and then selling two billion eggs a year—about 3 percent of all eggs in the United States: Strom, Stephanie, "McDonald's Plans a Shift to Eggs From Only Cage-Free Hens," *The New York Times*, September 9, 2015,http://www.nytimes.com/2015/09/10/business/McDonalds-to-use-eggs-from-only-cage-free-hens.html.

65 Taco Bell . . . announced it, too, would go 100 percent cage free: Gasparro, Annie, "Taco Bell to Switch to Cage-Free Eggs," *The Wall Street Journal*, November 16, 2015, http://www.wsj.com/articles/taco-bell-to-switch-to-cage-free-eggs-1447685367.

65 with six thousand outlets and an array of breakfast offerings that featured eggs: Gasparro, Annie, "Taco Bell to Switch to Cage-Free Eggs," *The Wall Street Journal*, November 16, 2015, http://www.wsj.com/articles/taco-bell-to-switch-to-cage-free-eggs-1447685367.

65 "Cage free is just the next logical step . . . It's the future of the industry.": Hickman's Family Farms, "Hickman's Family Farms Announces Major Cage-Free Egg-spansion: Egg Producer Converting to Cage-Free to Keep Up with Market Demands," n.d., http://www.hickmanseggs.com/.

65 "cage-free egg production will be the company's standard": Rembrandt Foods, "Cage-Free to Become the Standard for Rembrandt Foods," October 13, 2015,

http://www.rembrandtfoods.com/news-events/cage-free-to-become-the-standard-for-rembrandt-foods/.

65 "We welcome the growing movement of major food companies switching exclusively to cage-free eggs . . . with a reasonable time, we can meet any demand . . .", Rembrandt Foods, "Cage-Free to Become the Standard for Rembrandt Foods," October 13, 2015, http://www.rembrandtfoods.com/news-events/cage-free-to-become-the-standard-for-rembrandt-foods/.

66 Steve Herbruck . . . had moved four million of his seven million birds to cage-free houses by the time McDonald's made its announcement: Strom, Stephanie, "McDonald's Plans a Shift to Eggs From Only Cage-Free Hens," *The New York Times*, September 9, 2015, http://www.nytimes.com/2015/09/10/business/McDonalds-to-use-eggs-from-only-cage-free-hens.html.

66 Walmart's own customer surveys showed that 77 percent of its customers will "increase their trust" and 66 percent will "increase their likelihood to shop from a retailer that ensures humane treatment of livestock": D'Innocenzio, A., "Walmart Makes Public Guidelines to Suppliers on Antibiotics, Animal Treatment," *GlobalNews*, May 22, 2015, http://globalnews.ca/news/2011608/walmart-makes-public-guidelines-to-suppliers-on-antibiotics-animal-treatment/.

THREE: The Chicken or the Egg—or Neither?

69 "when you see how these cows are treated . . .": Jha, Alok, Synthetic Meat: How the World's Costliest Burger Made it on to the Plate," *The Guardian*, August 5, 2013, http://www.theguardian.com/science/2013/aug/05/synthetic-meat-burger-stem-cells.

69 "If what you're doing is not seen by some people as science fiction . . .": Jha, Alok, Synthetic Meat: How the World's Costliest Burger Made it on to the Plate," *The Guardian*, August 5, 2013, http://www.theguardian.com/science/2013/aug/05/synthetic-meat-burger-stem-cells.

70 the Environmental Protection Agency reports that the nation's 18,800 Confined Animal Feeding Operations (CAFOs): U.S. Environmental Protection Agency. National Pollutant Discharge Elimination System permit regulation and effluent limitation guidelines and standards for concentrated animal feeding operations (CAFOs); final rule. February 12, 2013. *Federal Register* 68(29): 7176, 7180.

71 A European Union study predicts lab-grown meat could reduce land use by 99.7 percent: Wakefield, Jane, "The Father and Son Planning Meat-Free Immortality," *BBC News*, November 12, 2015, http://www.bbc.com/news/technology-34581809.

71 "beef and pork went through a series of mutations . . .": Cronon, William. *Nature's Metropolis: Chicago and the Great West*, W.W. Norton & Company: New York, 225.

72 "2300 pens on a hundred acres, capable of handling 21,000 head of cattle . . .": Cronon, William. *Nature's Metropolis: Chicago and the Great West*, W.W. Norton & Company: New York, 210.

77 "countries with established histories of entomophagy . . . affordable and sustainable insect farming technologies . . . Not only do our durable farming units cre-

ate income stability for rural farmers . . .": Aspire, "Mission," n.d., http://www.aspirefg.com/mission/.

78 the company has already been valued at $100 million: Colt, Sam, "Soylent's Magical Milkshake Is Now Worth $100 Million," *Business Insider*, January 7, 2015, http://www.businessinsider.com/soylents-magical-milkshake-is-now-worth-100-million-2015-1.

78 The $60 billion leather industry processes twenty billion square feet of leather a year: United Nations Industrial Development Organization, *Future Trends in the World Leather and Leather Products Industry and Trade*, 2010, 28.

82 Since 1950, Americans have increased their consumption of meat and fish by almost 50 percent: USDA, "Profiling Food Consumption in America," *USDA Factbook*, 2010, 28, http://www.usda.gov/factbook/chapter2.pdf; USDA, Average Daily Intake of Food by Food Source and Demographic Characteristics (Data Set), http://www.ers.usda.gov/data-products/food-consumption-and-nutrient-intakes.aspx#26667.

82 Americans eat more meat per capita than just about any people in the world: FAO, "TABLE A3 Per capita consumption of livestock products, 1995–2005," *The State of Food and Agriculture*, 2009, 135–139, http://www.fao.org/docrep/012/i0680e/i0680e00.htm.

85 Ezra Taft Benson . . . famously declared, was "to get big or get out": Midkiff, Ken, "Introduction: Get Big or Get Out," *The Meat You Eat: How Corporate Farming Has Endangered America's Food Supply*, Macmillan, 2005, 3.

87 if humans grew as fast and as large as chickens, a 6.6 pound newborn baby would weigh 660 pounds after two months: Magdoff, Fred, "A Rational Agriculture is Incompatible with Capitalism," *Monthly Review*, March 2015, 10.

87 Cows at industrial dairies now yield an average of 27,000 pounds of milk a year: "University of New Hampshire Fairchild dairy produces 'gold' standard of milk," *Holstein World*, December 8, 2014.

87 a 91 percent loss in the number of pig producers: Leonard, Christopher, *The Meat Racket: The Secret Takeover of America's Food Business*, (Simon & Schuster, New York): 2014.

88 from about 1.5 billion animals in 1960 to 9 billion today: USDA, Economic Research Service, "Livestock and Meat Domestic Data, http://www.ers.usda.gov/data-products/livestock-meat-domestic-data.aspx.

90 Pinnacle Foods paid $174 million to acquire Garden Protein International: "Pinnacle Foods to Buy Canadian Protein Foods Maker Garden," Reuters, November 14, 2014, http://reut.rs/1EDXOzP.

90 2014 revenues of $4.1 billion: "Chipotle Mexican Grill, Inc. Announces Fourth Quarter and Full Year 2014 Results," Press Release, February 3, 2015, http://ir.chipotle.com/phoenix.zhtml?c=194775&p=irol-newsArticle&ID=2013178.

"Chipotle Mexican Grill Earnings Preview: All Eyes On Comparable Store Sales Growth," *Fortune*, April 20, 2015, http://www.forbes.com/sites/greatspeculations/2015/04/20/chipotle-mexican-grill-earnings-preview-all-eyes-on-comparable-store-sales-growth/.

91 In 2014, a study by the Humane Research Council revealed that 84 percent

of . . . . vegetarians later backtracked: Dahl, Melissa, "84 Percent of Vegetarians Go Back to Eating Meat," *New York Magazine*, December 3, 2014, http://nymag.com/scienceofus/2014/12/84-percent-of-vegetarians-go-back-to-eating-meat.html.

92 In July 2015, Google reportedly offered to buy Impossible Foods for $200–300 million: Millner, Jack, "Google Wants the World to Go Meat-Free: Search Giant Tried to Buy a Veggie Burger Start-Up for $300 Million," *The Daily Mail*, July 28, 2015, http://www.dailymail.co.uk/sciencetech/article-3177026/Google-wants-world-meat-free-Search-giant-tried-buy-veggie-burger-start-300-MILLION.html.

92 a second round of $108 million in capital announced in October 2015: Watson, Elaine, "Impossible Foods Raises $108 m; Prepares to Launch Plant-Based Burger by End of 2016," *Food Navigator USA*, October 8, 2015, http://www.foodnavigator-usa.com/R-D/Impossible-Foods-raises-108m-will-launch-plant-based-burger-in-2016.

92 "There are limits to what a cow can be . . .": Rusli, Evelyn, "So, What Does a Plant-Based Veggie Burger Taste Like?", *The Wall Street Journal*, October 7, 2014, http://blogs.wsj.com/digits/2014/10/07/taste-test-a-veggie-burger-that-looks-and-cooks-like-meat/.

92 In October 2015, Brown announced a new board member for his company: "A Former McDonald's CEO is Joining a Vegan Meat Startup," *Fortune*, November 9, 2015. http://fortune.com/2015/11/09/McDonalds-beyond-meat-thompson/.

95 A single egg producer . . . uses thirty million eggs a year for its own pasta product: Personal communication with Josh Balk, September 2015.

96 FDA cleared the way for the company to continue to use the term: Cowitt, Beth, "The Mayo Wars Just Ended," *Fortune*, December 17, 2015.

96 It had been in the crosshairs of . . . the American Egg Board: Thielman, S., and Rushe, D., "Government-Backed Egg Lobby Tried to Crack Food Startup, Emails Show," *The Guardian*, September 2, 2015, http://www.theguardian.com/us-news/2015/sep/02/usda-american-egg-board-hampton-creek-just-mayo.

96 emails discovered . . . the AEB conspired to thwart Hampton Creek's initial efforts to get its products placed in grocery stores: Thielman, S., and Rushe, D., "Government-Backed Egg Lobby Tried to Crack Food Startup, Emails, Show," *The Guardian*, September 2, 2015, http://www.theguardian.com/us-news/2015/sep/02/usda-american-egg-board-hampton-creek-just-mayo.

96 Joanne Ivy . . . suggested it "would be a good idea if Edelman looked at this product as a crisis and major threat to the future of the egg product business": Thielman, S., and Rushe, D., "Government-Backed Egg Lobby Tried to Crack Food Startup, Emails, Show," *The Guardian*, September 2, 2015, http://www.theguardian.com/us-news/2015/sep/02/usda-american-egg-board-hampton-creek-just-mayo.

97 "If these Great Depression era institutions have outlived their purpose, and if evidence suggests they behave like state-sponsored cartels that intimidate . . .": Romboy, Dennis, "Sen. Mike Lee Takes on Big Egg in Defense of Little Mayo," *Deseret News*, October 20, 2015, http://www.deseretnews.com/

article / 865639496 / Sen-Mike-Lee-takes-on-Big-Egg-in-defense-of-little-mayo.html ?pg=all.

100 "There is an inverse relationship between accelerated growth and disease resistance . . .": Entis, Laura, "Will the Worst Bird Flu Outbreak in US History Finally Make Us Reconsider Factory Farming Chicken?", *The Guardian*, July 14, 2015, http: / / www.theguardian.com / vital-signs / 2015 / jul / 14 / bird-flu -devastation-highlights-unsustainability-of-commercial-chicken-farming.

101 the US government had already handed over its first grant: Personal communication with Chad Gregory, September 2015.

101 the USDA in September 2015 issued a statement supporting in some cases shutting off the vents and heating up the houses: USDA, *HPAI Outbreak 2014–2015 Ventilation Shutdown Evidence & Policy*, September 18, 2015, https: / / www .aphis.usda.gov / animal_health / emergency_management / downloads / hpai / ventilationshutdownpolicy.pdf.

102 In June 2015, 7–Eleven . . . swapped out egg-based mayonnaise for Just Mayo: Kane, Colleen, "7-Eleven Quietly Switched to Vegan Mayo," *Fortune*, July 21, 2015, http: / / for.tn / 1CSDSyd.

102 "By 2050, the world's population will grow to more than nine billion . . .": Gates, Bill, "Future of Food," Gatesnotes, March 18, 2013, http: / / www.gatesnotes .com / About-Bill-Gates / Future-of-Food.

FOUR: **Now, That's Entertainment**

104 It was the fourth death of a real horse : Baum, Gary, "No Animals Were Harmed," *The Hollywood Reporter*, November 25, 2013, http: / / www.hollywoodreporter .com / feature / .

108 "creatures are coerced to perform unnatural and dangerous acts . . . ": "Inquiry into Use of Animals in Motion Picture Production Completed by Investigators," *Christian Science Monitor*, June 18, 1925.

109 More than one hundred horses perished in the making of *Ben Hur* in 1925: Baum, Gary, "No Animals Were Harmed," *The Hollywood Reporter*, November 25, 2013, http: / / www.hollywoodreporter.com / feature / .

109 Scandals surrounded the production of *The Charge of the Light Brigade* (1936): Flynn, Errol, *My Wicked, Wicked Ways: The Autobiography of Errol Flynn* (Cooper Square Press: New York), 2003, 212.

109 Howls were also raised about *Jesse James* (1939), whose director drove a horse off a cliff: Baum, Gary, "No Animals Were Harmed," *The Hollywood Reporter*, November 25, 2013, http: / / www.hollywoodreporter.com / feature / .

109 in the classic *Stagecoach*, director John Ford used trip wire to dramatically upend the horses: Mitchum, Petrine, *Hollywood Hoofbeats: Trails Blazed Across the Silver Screen* (BowTie Press: Irvine, CA), 2005.

109 "wires attached to the horses' forelegs . . .": Mitchum, Petrine, *Hollywood Hoofbeats: Trails Blazed Across the Silver Screen* (BowTie Press: Irvine, CA), 2005.

109 in the 1966 Soviet-era classic film *Andrea Rublev*, a horse is pushed down a flight of stairs: Johnson, Vida, and Graham Petrie, "Andrei Rublev," *The Films of Andrei Tarkovsky: A Visual Fugue* (Indiana University Press: Bloomington, IN), 1994, 94.

109 In 1903 . . . paying customers watched two hucksters poison, strangle, torment, and electrocute an Asian Elephant: St. Clair, Jeffery, "Let Us Now Praise Infamous Animals," *Fear of the Animal Planet: The Hidden History of Animal Resistance* (AK Press: Oakland, CA), 2011.

109 Topsy . . . killed a man who had been taunting her: Daly, Michael, "Topsy and the Tormenter," *Topsy: The Startling Story of the Crooked-Tailed Elephant, P. T. Barnum, and the American Wizard, Thomas Edison* (Grove Press: New York, NY), 2014, 302.

110 Edison Electric recorded *Electrocuting the Elephant*: "Lynching Photography at the Turn of the Nineteenth Century," *A Spectacular Secret: Lynching in American Life and Literature* (University of Chicago Press: Chicago, CA), 2006, 225.

110 Lemmings don't naturally engage in mass suicidal behavior—the animals were physically thrown into the water: Woodford, Riley, "Lemming Suicide Myth: Disney Film Faked Bogus Behavior," *Alaska Fish & Wildlife News*, September 2003, http://www.adfg.alaska.gov/index.cfm?adfg=wildlifenews.view_article&articles_id=56.

110 According to Jane Goodall, Clyde was trained with a can of mace and a pipe wrapped in newspaper: Abramowitz, Rachel, " 'Every Which Way but Abuse' Should be Motto," *The Los Angeles Times*, August 27, 2008, http://articles.latimes.com/2008/aug/27/entertainment/et-brief27.

110 "Near the end of filming . . . He died soon after . . .", Abramowitz, Rachel, " 'Every Which Way but Abuse' Should be Motto," *The Los Angeles Times*, August 27, 2008, http://articles.latimes.com/2008/aug/27/entertainment/et-brief27.

111 The film . . . produced a number of dead horses, and the director even staged live cockfights: American Humane Association, "Heaven's Gate," n.d., http://www.humanehollywood.org/index.php/movie-archive/item/heaven-s-gate.

111 "I've watched the AHA a long time . . .": Smith, Lucinda, Leah Feldon, and Eleanor Hoover, "Speaking Up for 'Abused' Animals, Bob Barker is Hit with a Lawsuit," *People*, September 18, 1989, http://www.people.com/people/archive/article/0,,20121207,00.html.

111 Derby and television icon and animal advocate Bob Barker have been among the most outspoken critics of the organization: Smith, Lucinda, Leah Feldon, and Eleanor Hoover, "Speaking Up for 'Abused' Animals, Bob Barker is Hit with a Lawsuit," *People*, September 18, 1989, http://www.people.com/people/archive/article/0,,20121207,00.html.

112 "This one take with him just went really bad . . .": Baum, Gary, "No Animals Were Harmed," *The Hollywood Reporter*, November 25, 2013, http://www.hollywoodreporter.com/feature/.

116 "There are endless possibilities . . . more motorsports, daredevils, and feats of human physical capabilities": *CBS News* and Associated Press, "Ringling Bros. Eliminating Elephant Acts Amid Public 'Mood Shift,' " March 5, 2015, http://www.cbsnews.com/news/ringling-bros-eliminating-elephant-acts-amid-public-mood-shift/.

116 "We're always changing and we're always learning . . .": *CBS News* and Associated Press, "Ringling Bros. Eliminating Elephant Acts Amid Public 'Mood Shift,' " March 5, 2015, http://www.cbsnews.com/news/ringling-bros-eliminating-elephant-acts-amid-public-mood-shift/.

116 "I don't know any 145-year-old company . . .": Claman, Liz, "Feld CEO: Ringling Bros' Elephant Phase-Out About Company Survival," *Business Insiders*, March 5, 2015, http://www.foxbusiness.com/business-leaders/2015/03/05/feld-ceo-ringling-bros-elephant-phase-out-about-company-survival/.

117 In 1906, the Bronx Zoo confined a Congolese pygmy named Ota Benga: Keller, Mitch, "The Scandal at the Zoo," *The New York Times*, August 6, 2006, http://www.nytimes.com/2006/08/06/nyregion/thecity/06zoo.html?pagewanted=all.

117 protests ensued over the dehumanizing treatment of Benga: Newkirk, Pamela, "The man who was caged in a zoo," *The Guardian*, June 3, 2015. http://www.theguardian.com/world/2015/jun/03/the-man-who-was-caged-in-a-zoo.

117 In the 1920s . . . Ringling Bros. cancelled its large animal acts in the second half of the decade: Unti, Bernard. "The Quality of Mercy: Organized Animal Protection in the United States, 1866–1930." Ph.D. diss., American University, 2002; and Davis, Janet M. The *Gospel of Kindness: Animal Welfare and the Making of Modern America*. New York: Oxford University Press, 2016.

118 "We have detected a shift in mood . . .": Izadi, Elahe, "The Long Battle to Remove Elephants From the Ringling Bros. Circus," *The Washington Post*, March 5, 2015, http://www.washingtonpost.com/news/morning-mix/wp/2015/03/05/the-long-battle-to-remove-elephants-from-the-ringling-bros-circus/.

118 Mr. Feld also said it had been hard for Ringling: Claman, Liz, "Feld CEO: Ringling Bros. Elephant Phase-Out About Company Survival," *Business Insiders*, March 5, 2015, http://www.foxbusiness.com/business-leaders/2015/03/05/feld-ceo-ringling-bros-elephant-phase-out-about-company-survival/.

118 Los Angeles and Oakland . . . had banned bull hooks: Izadi, Elahe, "The Long Battle to Remove Elephants From the Ringling Bros. Circus," *The Washington Post*, March 5, 2015, http://www.washingtonpost.com/news/morning-mix/wp/2015/03/05/the-long-battle-to-remove-elephants-from-the-ringling-bros-circus/.

118 "I can't fight city hall": Ager, Susan, "Here's Where Ringling Bros. Is Sending Its Circus Elephants to Retire," *National Geographic*, September 17, 2015, http://news.nationalgeographic.com/2015/09/150916-ringling-circus-elephants-florida-center/.

119 Over the years, Ringling's Ken Feld earned a reputation for ruthlessness and cunning: Stein, Jeff, "The Greatest Vendetta on Earth," *Salon*, August 30, 2001, http://www.salon.com/2001/08/30/circus_2/.

119 Feld hired Clair George, a former deputy director of operations for the Central Intelligence Agency: Leung, Rebecca, "Exclusive: Send in the Spies?", *60 Minutes*, May 1, 2003, http://www.cbsnews.com/news/exclusive-send-in-the-spies/.

119 When Pottker signaled that she wanted to dig in deeper: Stein, Jeff, "The Greatest Vendetta on Earth," *Salon*, August 30, 2001, http://www.salon.com/2001/08/30/circus_2/.

119 George in turn enlisted an operative: Leung, Rebecca, "Exclusive: Send in the Spies?", *60 Minutes*, May 1, 2003, http://www.cbsnews.com/news/exclusive-send-in-the-spies/.

119 This intrigue . . . was an eight-year gambit for Feld: Leung, Rebecca, "Exclusive: Send in the Spies?", *60 Minutes*, May 1, 2003, http://www.cbsnews.com/news/exclusive-send-in-the-spies/.

119 Pottker estimates that Feld spent $3 million: Leung, Rebecca, "Exclusive: Send in the Spies?", *60 Minutes*, May 1, 2003, http://www.cbsnews.com/news/exclusive-send-in-the-spies/.

119 George, who had been drummed out of the CIA after his involvement: Leung, Rebecca, "Exclusive: Send in the Spies?", *60 Minutes*, May 1, 2003, http://www.cbsnews.com/news/exclusive-send-in-the-spies/.

119 allegedly oversaw the infiltration by Feld operatives of at least two animal-protection groups: Stein, Jeff, "The Greatest Vendetta on Earth," *Salon*, August 30, 2001, http://www.salon.com/2001/08/30/circus_2/.

119 Feld . . . ultimately paying PAWS an undisclosed financial settlement: Stein, Jeff, "The Greatest Vendetta on Earth," *Salon*, August 30, 2001, http://www.salon.com/2001/08/30/circus_2/.

120 Feld became embroiled in a legal bout with another set of animal welfare groups: Nelson, Deborah, "The Cruelest Show on Earth," *Mother Jones*, November/December 2011, http://www.motherjones.com/environment/2011/10/ringling-bros-elephant-abuse?page=1.

120 That case . . . lasted all of fourteen years: Heath, Thomas, "Ringling Circus Prevails in 14-Year Legal Case; Collects $16M from Humane Society, Others," *The Washington Post*, May 16, 2014, http://wapo.st/1gaY2Ih.

120 US District Judge Emmet Sullivan denied the animal-protection groups standing: Nelson, Deborah, "The Cruelest Show on Earth," *Mother Jones*, November/December 2011, http://www.motherjones.com/environment/2011/10/ringling-bros-elephant-abuse?page=1.

120 Feld . . . went on the offense, filing a civil RICO case: Heath, Thomas, "Ringling Circus Prevails in 14-Year Legal Case; Collects $16M from Humane Society, Others," *The Washington Post*, May 16, 2014, http://wapo.st/1gaY2Ih.

120 The legal saga . . . was settled just 10 months before Ringling's announcement to retire the elephants: Pérez-Peña, Richard, "Elephants to Retire From Ringling Brothers Stage," *The New York Times*, March 5, 2015, http://www.nytimes.com/2015/03/06/us/ringling-brothers-circus-dropping-elephants-from-act.html?_r=0.

120 The animal groups . . . ended up paying Feld's legal fees: Heath, Thomas, "Ringling Circus Prevails in 14-Year Legal Case; Collects $16M from Humane Society, Others," *The Washington Post*, May 16, 2014, http://wapo.st/1gaY2Ih.

120 In November 2011, the US Department of Agriculture fined Ringling $270,000: Sacks, David, and Lyndsay Cole, "USDA and Feld Entertainment, Inc., Reach Settlement Agreement," News Release, 11/28/2011, http://www.usda.gov/wps/portal/usda/usdahome?contentid=2011/11/0494.xml.

120 numerous alleged violations of the Animal Welfare Act: Simpson, Ian, "Ringling Circus Pays Record Fine in Animal Welfare Case," Reuters, http://www.reuters.com/article/2011/11/29/us-usda-ringling-idUSTRE7AS2AG20111129.

120 The fine against Ringling . . . was nonetheless the largest animal-welfare pen-

alty: Simpson, Ian, "Ringling Circus Pays Record Fine in Animal Welfare Case," Reuters, http://www.reuters.com/article/2011/11/29/us-usda-ringling-idUSTRE7AS2AG20111129.

121 Feld Entertainment . . . now brought in $1 billion every year: Mac, Ryan, "Ringling Bros. Owner Not Clowning Around with Business, Cannons to Billionaire Status," *Forbes*, January 28, 2014, http://www.forbes.com/sites/ryanmac/2014/01/28/ringling-bros-owner-not-clowning-around-with-business-cannons-to-billionaire-status/.

123 Janet . . . went on a rampage in 1992: Sahagun, Louis, "Elephants Pose Giant Danger: Pachyderm Attacks are on the Rise, Creating King-Size Problems for Zoos and Circuses. Some Believe the Intelligent Mammals are Rebelling Against Inhumane Treatment," *Los Angeles Times*, October 11, 1994, http://articles.latimes.com/1994–10–11/news/mn-49101_1_captive-elephant.

123 Doyle . . . has been active ever since as an advocate: Sahagun, Louis, "Elephants Pose Giant Danger: Pachyderm Attacks are on the Rise, Creating King-Size Problems for Zoos and Circuses. Some Believe the Intelligent Mammals are Rebelling Against Inhumane Treatment," *Los Angeles Times*, October 11, 1994, http://articles.latimes.com/1994–10–11/news/mn-49101_1_captive-elephant.

124 Tyke attacked a circus groom . . .: Sahagun, Louis, "Elephants Pose Giant Danger: Pachyderm Attacks are on the Rise, Creating King-Size Problems for Zoos and Circuses. Some Believe the Intelligent Mammals are Rebelling Against Inhumane Treatment," *Los Angeles Times*, October 11, 1994, http://articles.latimes.com/1994–10–11/news/mn-49101_1_captive-elephant.

124 Tyke had been at the center of two other uncontrolled rampages . . .: Sahagun, Louis, "Elephants Pose Giant Danger: Pachyderm Attacks are on the Rise, Creating King-Size Problems for Zoos and Circuses. Some Believe the Intelligent Mammals are Rebelling Against Inhumane Treatment," *Los Angeles Times*, October 11, 1994, http://articles.latimes.com/1994–10–11/news/mn-49101_1_captive-elephant.

125 "They are required to perform behaviors never seen in nature . . . In short they are treated as commodities . . .": Moss, Cynthia, Amboseli Trust for Elephants, ATE Statement on Circus Elephants, November 20, 2014. https://www.elephanttrust.org/index.php/articles/item/this-is-the-article-title-copy-2.

125 Robert Ridley . . . testified that he saw "puncture wounds . . . hook boils": United States District Court for the District of Columbia, "Plaintiffs' Proposed Findings of Fact," *American Society for the Prevention of Cruelty to Animals, et al., v. Feld Entertainment, Inc.*, 2009, 65, https://www.animallaw.info/sites/default/files/pbusfdaspca_ringlingbros_plaintiffs_findings_facts.pdf.

125 One Ringling Memorandum reported that an elephant had sustained twenty-two puncture wounds: United States District Court for the District of Columbia, "Plaintiffs' Proposed Findings of Fact," *American Society for the Prevention of Cruelty to Animals, et al., v. Feld Entertainment, Inc.*, 2009, 65, https://www.animallaw.info/sites/default/files/pbusfdaspca_ringlingbros_plaintiffs_findings_facts.pdf.

125 A separate internal memorandum . . . reported that an elephant was bleeding: United States District Court for the District of Columbia, "Plaintiffs' Proposed

Findings of Fact," *American Society for the Prevention of Cruelty to Animals, et al., v. Feld Entertainment, Inc.*, 2009, 65, https://www.animallaw.info/sites/default/files/pbusfdaspca_ringlingbros_plaintiffs_findings_facts.pdf.

125 According to Ringling Bros. emails: United States District Court for the District of Columbia, "Plaintiffs' Proposed Findings of Fact," *American Society for the Prevention of Cruelty to Animals, et al., v. Feld Entertainment, Inc.*, 2009, 222, https://www.animallaw.info/sites/default/files/pbusfdaspca_ringlingbros_plaintiffs_findings_facts.pdf.

125 Kenneth Feld himself testified: United States District Court for the District of Columbia, "Plaintiffs' Proposed Findings of Fact," *American Society for the Prevention of Cruelty to Animals, et al., v. Feld Entertainment, Inc.*, 2009, 96, https://www.animallaw.info/sites/default/files/pbusfdaspca_ringlingbros_plaintiffs_findings_facts.pdf.

126 Ringling's performing elephants might go to as many as 115 cities a year: Associated Press, "Ringling Bros.' Circus to End Use of Elephants," *The Wall Street Journal*, March 6, 2015, http://www.wsj.com/articles/ringling-bros-circus-to-end-use-of-elephants-1425570518.

126 Ringling's own "transportation orders": Nelson, Deborah, "The Cruelest Show on Earth," *Mother Jones*, November/December 2011, http://www.motherjones.com/environment/2011/10/ringling-bros-elephant-abuse?page=1.

126 In the past ten years tuberculosis has been diagnosed in as many as nineteen of Ringling's fifty or so elephants: Goldstein, Sasha, "Ringling Bros. and Barnum & Bailey Elephant Phase Out Due to Rampant Tuberculosis: PETA," *New York Daily News*, March 6, 2015, http://www.nydailynews.com/news/national/ringling-bros-elephant-retirement-due-rampant-tb-peta-article-1.2140041.

127 "terrific news for a majestic species of animal . . . a triumph of public awareness": Editorial Board, "Ringling's Elephant Move is a Start," *The Boston Globe*, March 6, 2015, https://www.bostonglobe.com/opinion/editorials/2015/03/05/ringling-elephant-move-start/9al0WFYVItuOsqKNTflzWN/story.html.

127 "these magnificent creatures deserve better . . .": Editorial Board, "Why Not Retire the Circus Elephants Now?", *The New York Times*, March 6, 2015, http://www.nytimes.com/2015/03/07/opinion/why-not-retire-the-circus-elephants-now.html?_r=0.

127 "'The Greatest Show on Earth' is soon to be even greater . . .": Editorial Board, "Ringling Bros. Cleaning Up its Elephant Act," *The Dallas Morning News*, March 6, 2015, http://www.dallasnews.com/opinion/editorials/20150306-editorial-hits-and-misses.ece.

127 "represents the evolution from the old circus to the new": Editorial Board, "Circus Takes a Big Step Forward," *The Sarasota Herald-Tribune*, March 10, 2015, http://www.heraldtribune.com/article/20150311/opinion/303119996?p=1&tc=pg.

127 "If you are going to do more shows . . .": Enquirer Editorial Board, "Can You Ask Tougher Questions than a Third-Grader?", *The Cincinnati Enquirer*, May 6, 2015, http://www.cincinnati.com/story/opinion/editorials/2015/05/05/can-ask-tougher-questions-third-grader/26922415/.

127 Cunningham explained that having elephants perform tricks "doesn't appeal to our higher selves . . . There are so many different types of entertainment":

Cronin, Melissa, "International Circus Announces End to 'Outdated' Elephant Act," The Dodo, April 13, 2015, https://www.thedodo.com/circus-elephants-carden-1089266986.html.

128 a number of circuses gave up their animal acts: PETA, "Animal-Free Circuses," Factsheet, n.d., http://www.mediapeta.com/peta/pdf/animal-free-circuses-pdf.pdf.

128 More than two dozen nations had banned the use of elephants for entertainment: Casamitjana, Jordi, "Which countries have best/worst records on banning animals from circuses?," International Fund for Animal Welfare. http://www.ifaw.org/international/node/104875.

128 In Australia and Canada, by 2014, reformers had secured bans: Stop Circus Suffering, "Circus Bans," n.d., http://www.stopcircussuffering.com/circus-bans/.

128 Governor David Ige of Hawaii pledged: Lang, Davi, "Hawaii Poised to Become the First State to Ban Wild Animal Entertainment Acts," Winning the Cast Against Cruelty, May 12, 2015, http://aldf.org/blog/hawaii-poised-to-become-the-first-state-to-ban-wild-animal-entertainment-acts/.

128 California lawmakers banned bull hooks: Koseff, Alexei, "Which Bills Will Jerry Brown Sign?" *The Sacramento Bee*, September 17, 2015, http://www.sacbee.com/news/politics-government/capitol-alert/article35665722.html.

128 launched Cirque du Soleil in 1984 as a "group of stilt-walkers and street performers" . . . generating $1 billion in revenue: Gittleson, Kim, "How Cirque du Soleil Became a Billion Dollar Business," *BBC News*, December 12, 2013, http://www.bbc.com/news/business-25311503.

129 Guy Laliberté, who sold a majority stake in the company for $1.5 billion: Hunter-Tilney, Lucovic, "Guy Laliberté, Cirque du Soleil Co-Founder," *Financial Times*, April 24, 2015, http://www.ft.com/cms/s/0/5156e710-e99a-11e4-a687–00144feab7de.html#axzz3mUOGChPg.

129 "festooning these magnificent creatures . . .": Krauthammer, Charles, "Free Willy!", *The Washington Post*, May 7, 2015, https://www.washingtonpost.com/opinions/free-willy/2015/05/07/4d1a82f2-f4f2–11e4-b2f3-af5479e6bbdd_story.html.

130 SeaWorld . . . collectively drawing twenty-four million patrons a year: CBS, "Will SeaWorld's Killer Whale Trainers Swim with Orcas Again? Federal Judges to Decide," *CBS This Morning*, November 13, 2013, http://www.cbsnews.com/news/will-seaworlds-killer-whale-trainers-swim-with-orcas-again-federal-judges-to-decide/.

131 Cowperthwaite drew twenty-one million television viewers in one week: Hewitt, Michael, " 'Blackfish,' 'The Jinx' Lead Trend of Documentaries Hitting it Big on TV," *The Orange County Register*, June 29, 2015, http://www.ocregister.com/articles/hbo-668989-documentaries-film.html.

132 SeaWorld's own CEO quietly sold $3 million of his SeaWorld shares: Richardson, Matthew, "SeaWorld CEO Sells 100,000 Shares and Rakes in Millions," *Orlando Business Journal*, February 6, 2014, http://www.bizjournals.com/orlando/morning_call/2014/02/seaworld-ceo-sells-100000-shares-and.html.

132 "I don't agree with the way they treat their animals": Duke, Alan, "Pat Benatar,

Beach Boys Join 'Blackfish' Cancellation List," *CNN*, January 16, 2014, http://www.cnn.com/2014/01/16/showbiz/blackfish-busch-gardens-cancellations/.

132 SeaWorld gave notice to more than three hundred employees: Pedicini, Sandra, "SeaWorld Lays Off Clyde & Seamore Mimes," *Orlando Sentinel*, December 16, 2014, http://www.orlandosentinel.com/business/os-seaworld-lays-off-mimes-20141216-story.html.

132 including its CEO: Bilbao, Richard, "SeaWorld CEO Jim Atchison to Step Down, Layoffs Coming," *Orlando Business Journal*, December 11, 2014, http://www.bizjournals.com/orlando/blog/2014/12/seaworld-ceo-jim-atchison-to-step-down-layoffs.html.

133 with revenues up from $100 million in 1990 to $1.4 billion in 2012: Garcia, Jason, "SeaWorld Entertainment's Profit Soars on Increases in Revenue, Attendance," *Orlando Sentinel*, March 26, 2013, http://articles.orlandosentinel.com/2013–03–26/business/os-seaworld-profits-triple-20130326_1_profit-soars-seaworld-orlando-seaworld-san-antonio.

133 *Free Willy*, which grossed more than $150 million: "Free Willy," Box Office Mojo, n.d., http://www.boxofficemojo.com/movies/?id=freewilly.htm.

133 Philanthropist and newspaper publisher Wendy McCaw pitched in with . . . the producers of *Free Willy*, and thousands of school children in putting up $7 million: Dornin, Rusty, " 'Free Willy' Star at Home in Oregon," CNN, January 8, 1996, http://www.cnn.com/WORLD/9601/keiko_willy_moves/01–08/.

135 He'd killed a female trainer . . . in 1991 at a makeshift marine park in Victoria, British Columbia: "Killer Whale Kills SeaWorld Trainer, *CBC News*, February 24, 2010, http://www.cbc.ca/news/world/killer-whale-kills-seaworld-trainer-1.892016.

135 operators shuttered the place and put the facility's three killer whales on the open market: Kuo, Vivian, "Orca Trainer Saw Best of Keiko, Worst of Tilikum," CNN, October 26, 2013, http://www.cnn.com/2013/10/26/world/americas/orca-trainer-tilikum-keiko/.

135 SeaWorld bought Tilikum and flew him to Orlando: Kuo, Vivian, "Orca Trainer Saw Best of Keiko, Worst of Tilikum," CNN, October 26, 2013, http://www.cnn.com/2013/10/26/world/americas/orca-trainer-tilikum-keiko/.

135 Tilikum apparently killed twenty-seven-year-old Daniel Dukes: "Killer Whale Kills SeaWorld Trainer, *CBC News*, February 24, 2010, http://www.cbc.ca/news/world/killer-whale-kills-seaworld-trainer-1.892016.

136 SeaWorld . . . attempted to frame the tragedy: Garcia, Jason, "Blackstone Chief Blames Brancheau for Own Death, Contradicting SeaWorld," *Orlando Sentinel*, January 24, 2014, http://www.orlandosentinel.com/business/tourism/tourism-central-florida-blog/os-blackstone-chief-blames-brancheau-for-own-death-contradicting-seaworld-20140124-post.html.

136 "The only thing that led to this event was a mistake made by Ms. Brancheau . . .": "Decision and Order," *Secretary of Labor, v. SeaWorld of Florida, LLC*, 2012, http://www.oshrc.gov/decisions/html_2012/10–1705.htm.

136 OSHA came to a different conclusion: "Decision and Order," *Secretary of Labor, v. SeaWorld of Florida, LLC*, 2012, http://www.oshrc.gov/decisions/html_2012/10–1705.htm.

136 In August 2010, the federal agency issued a "willful" citation: OSHA, "US Labor Department's OSHA cites SeaWorld of Florida following animal trainer's death," *News Release*, August 23, 2010, https://www.osha.gov/pls/oshaweb/owadisp.show_document?p_table EWS_RELEASES&p_id=18207.

137 He concluded that SeaWorld believed it was doing enough: "Decision and Order," *Secretary of Labor, v. SeaWorld of Florida, LLC*, 2012, http://www.oshrc.gov/decisions/html_2012/10–1705.htm.

137 The judge determined that Tilikum had grabbed Brancheau by the arm: "Decision and Order," *Secretary of Labor, v. SeaWorld of Florida, LLC*, 2012, http://www.oshrc.gov/decisions/html_2012/10–1705.htm.

137 Yet as late as 2014, the chief executive of Blackstone told CNBC that the deceased trainer "violated all the safety rules that we had": Garcia, Jason, "Blackstone Chief Blames Brancheau for Own Death, Contradicting SeaWorld," *Orlando Sentinel*, January 24, 2014, http://www.orlandosentinel.com/business/tourism/tourism-central-florida-blog/os-blackstone-chief-blames-brancheau-for-own-death-contradicting-seaworld-20140124-post.html.

137 the judge finding that SeaWorld knew from a series of accidents: "Decision and Order," *Secretary of Labor, v. SeaWorld of Florida, LLC*, 2012, http://www.oshrc.gov/decisions/html_2012/10–1705.htm.

137 He upheld OSHA's core claim: "Decision and Order," *Secretary of Labor, v. SeaWorld of Florida, LLC*, 2012, http://www.oshrc.gov/decisions/html_2012/10–1705.htm.

137 SeaWorld responded to the ruling: Zajac, Andrew, and Christopher Palmeri, "SeaWorld Trainers Barred From Killer Whale Pools, Rides," *Bloomberg Business*, April 11, 2014, http://www.bloomberg.com/news/articles/2014–04–11/seaworld-ban-on-killer-whale-contact-upheld-by-court.

137 The appellate court in April 2014 found: Zajac, Andrew, and Christopher Palmeri, "SeaWorld Trainers Barred From Killer Whale Pools, Rides," *Bloomberg Business*, April 11, 2014, http://www.bloomberg.com/news/articles/2014–04–11/seaworld-ban-on-killer-whale-contact-upheld-by-court.

137 In May 2015 . . . the California Division of Occupational Safety (CDOS) fined SeaWorld $25,770: Pedicini, Sandra, "Report: SeaWorld San Diego Fined for Safety Violations," *Orlando Sentinel*, May 1, 2015, http://www.orlandosentinel.com/business/os-seaworld-fined-safety-20150501-story.html.

137 CDOS cited SeaWorld for its failure to protect employees: Miller, Michael, "SeaWorld Fined for Improperly Protecting Employees from Killer Whales," *The Washington Post*, May 1, 2015, http://www.washingtonpost.com/news/morning-mix/wp/2015/05/01/seaworld-fined-for-improperly-protecting-employees-from-killer-whales/.

138 the State of California criticized SeaWorld: Miller, Michael, "SeaWorld Fined for Improperly Protecting Employees from Killer Whales," *The Washington Post*, May 1, 2015, http://www.washingtonpost.com/news/morning-mix/wp/2015/05/01/seaworld-fined-for-improperly-protecting-employees-from-killer-whales/.

142 "The whole family is out here twenty-five yards away maybe . . .": DiFusco, Danielle, "The Future of SeaWorld: What is Next for the Aquatic Tycoon; Can

They Survive the Backlash of 'Blackfish'?", *Prezi*, October 8, 2014, https://prezi.com/0kgtc_sr2qkh/the-future-of-seaworld-can-this-aquatic-tycoon-survive-the/.

143 the Georgia Aquarium . . . attempting in 2012 to gain permission to import eighteen beluga whales: NOAA Fisheries, *Georgia Aquarium Application to Import 18 Beluga Whales (File No. 17324)*, September 29, 2015, http://www.nmfs.noaa.gov/pr/permits/georgia_aquarium_belugas.htm.

143 The US government . . . stopped the transfer: NOAA Fisheries, *Georgia Aquarium Application to Import 18 Beluga Whales (File No. 17324)*, September 29, 2015, http://www.nmfs.noaa.gov/pr/permits/georgia_aquarium_belugas.htm.

143 a US District Court upheld the . . . denial of an import permit application: NOAA Fisheries, *Georgia Aquarium Application to Import 18 Beluga Whales (File No. 17324)*, September 29, 2015, http://www.nmfs.noaa.gov/pr/permits/georgia_aquarium_belugas.htm.

145 Manby . . . launched a $10 million advertising and social media campaign: Titlow, John, "SeaWorld is Spending $10 Million to Make You Forget About 'Blackfish,'" *Fast Company*, August 4, 2015, http://www.fastcompany.com/3046342/seaworld-is-spending-10-million-to-make-you-forget-about-blackfish.

145 In July 2015 . . . SeaWorld announced its quarterly earnings: SeaWorld Entertainment, "SeaWorld Entertainment, Inc. Reports Second Quarter 2015 Results," Q2 2015 Earnings Release, August 6, 2015, http://www.seaworldinvestors.com/files/doc_news/2015-Q2-Earnings-Release-IR-Website.pdf.

148 In 2014, scientists found a 104-year-old killer whale off the western coast of Canada: Schweig, Sarah, "104-Year-Old 'Granny' Orca Spotted Leaping Wildly Out Of The Ocean," bites @ Animal Planet, July 23, 2014, http://blogs.discovery.com/bites-animal-planet/2015/07/104-year-old-granny-orca-spotted-leaping-wildly-out-of-the-ocean.html.

148 One of her grandchildren, Canuck, reportedly died at the age of four after being captured and held at SeaWorld: Krushel, Jacob, "62 Orcas Have Died At SeaWorld—Not A Single One From Old Age," The Dodo, June 6, 2014, https://www.thedodo.com/62-orcas-have-died-at-seaworld-580775893.html.

148 The Whale and Dolphin Conservation project estimates that orcas born in captivity only live to four-and-a-half years old, on average; many of the long-living orcas at SeaWorld don't make it out of their twenties: WDC, "The Fate of Captive Orcas," n.d., http://us.whales.org/wdc-in-action/fate-of-captive-orcas

FIVE: **Animal Testing Yields to Humane Science**

151 "undermines the scientific, welfare, and conservation goals": Lincoln Park Zoo Press Release, "Lincoln Park Zoo Study Reveals That Inappropriate Media Portrayal of Chimpanzees May Hinder Conservation Efforts," March 13, 2008.

155 It costs roughly $22,000 a year to care for a chimp in a lab: Personal communication with Kathleen Conlee, November 2015. Also see "Costs for Maintaining Humane Care and Welfare of Chimpanzees," National Institutes of Health, Office of Research Infrastructure Programs, October 21, 2015.

157 One laboratory director . . . even compared these chimps to books in a library: Bonar, Laura, "20 NM Chimps Stuck in a Texas Lab Must Be Freed," *Albuquerque*

Journal, June 22, 2015, http://www.abqjournal.com/601986/opinion/20-nm-chimps-stuck-in-a-texas-lab-must-be-freed.html.

158 "unless these deficiencies are corrected, we would consider future studies conducted at your facility to be seriously flawed." Roller, Harriette, "Lethal Kinship: A Report on the Chimpanzees of the Coulston Foundation," www.apnm.org/campaigns/chimps/LethalKinship.doc.

160 750 or so remain in laboratories . . . including 200 chimps that the government seized from Coulston: "U.S. puts an end to all experiments on chimps," Mother Nature Network, November 19, 2015.

160 "I think it is important to find solutions to hepatitis C, but there are other ways to do it rather than testing on chimpanzees": KOAT 7, "Richardson Meets with NIH Over Alamogordo Chimps," August 17, 2010, http://www.koat.com/Richardson-Meets-With-NIH-Over-Alamogordo-Chimps/6136886.

160 "Considering the great progress the scientific community has made in research techniques . . .": Arcus Foundation, "Update: Transfer of Alamagordo Chimps on Hold," January 7, 2011, http://www.arcusfoundation.org/update-transfer-of-alamagordo-chimps-on-hold/.

161 Collins . . . postponed transferring the chimps to Texas: Arcus Foundation, "Update: Transfer of Alamagordo Chimps on Hold," January 7, 2011, http://www.arcusfoundation.org/update-transfer-of-alamagordo-chimps-on-hold/.

161 "any analysis of necessity must take these ethical issues into account": de Waal, Frans B.M., "Research Chimpanzees May Get a Break," *PLoS Biology*, 2012, 10(3), http://www.ncbi.nlm.nih.gov/pmc/articles/PMC3313912/.

161 "the chimpanzee's genetic proximity to humans . . .": Altevogt, B.M., D.E. Pankevich, M.K. Shelton-Davenport, et al., editors, "Conclusions and Recommendations," *Chimpanzees in Biomedical and Behavioral Research: Assessing the Necessity*, 2011, http://www.ncbi.nlm.nih.gov/books/NBK91452/.

161 "most current use of chimpanzees for biomedical research is unnecessary": Altevogt B.M., D.E. Pankevich , M.K. Shelton-Davenport, et al., editors, "Conclusions and Recommendations," *Chimpanzees in Biomedical and Behavioral Research: Assessing the Necessity*, 2011, http://www.ncbi.nlm.nih.gov/books/NBK91452/.

161 "the bar is very high" for any future uses of these animals in research: *CBS News* Staff, "Chimp Research Restrictions Underway, Gov't Says," December 16, 2011, http://www.cbsnews.com/news/chimp-research-restrictions-underway-govt-says/.

161 "very compelling and scientifically rigorous": Voelker, Rebecca, "Chimpanzees Aren't Necessary for Most Research, Says IOM," News@JAMA, December 15, 2011, http://newsatjama.jama.com/2011/12/15/chimpanzees-arent-necessary-for-most-research-says-iom/.

161 "decided to accept the IOM committee recommendations": Statement by NIH Director Dr. Francis Collins on the Institute of Medicine Report Addressing the Scientific Need for the Use of Chimpanzees in Research, December 15, 2011, http://www.nih.gov/news-events/news-releases/statement-nih-director-dr-francis-collins-institute-medicine-report-addressing-scientific-need-use-chimpanzees-research.

162 "greatly reducing their use in biomedical research is scientifically sound and the right thing to do": National Institutes of Health, NIH to Reduce Significantly the Use of Chimpanzees in Research, June 26, 2013, http://www.nih.gov/news-events/news-releases/nih-reduce-significantly-use-chimpanzees-research.

162 "recent advances in alternate research tools have rendered chimpanzees largely unnecessary as research subjects": Chimpanzees in Biomedical and Behavioral Research: Assessing the Necessity, December 2011, https://iom.nationalacademies.org/~/media/Files/Report%20Files/2011/Chimpanzees/chimpanzeereportbrief.pdf.

162 "I think that it's going to raise the bar for non-NIH-funded research . . .": Wadman, Meredith, "US Chimpanzee Research to be Curtailed," *Nature,* December 16, 2011, http://www.nature.com/news/us-chimpanzee-research-to-be-curtailed-1.9663.

163 "fully supports the findings of the 2011 Institute of Medicine report": Abbott Pharmaceuticals, "Animal Welfare Policy," http://www.abbott.com/policies/animal-welfare.html.

163 "pretty much everybody has gotten out, or is getting out, of research with chimps": Radnofsky, Louise, and Thomas M. Burton, "New Federal Rule Classifies Captive Chimpanzees as Endangered," *The Wall Street Journal,* June 12, 2015, http://www.wsj.com/articles/new-federal-rule-classifies-captive-chimpanzees-as-endangered-1434143642.

164 blowout at the Deepwater Horizon rig killed eleven workers: "Gulf Oil Spill Deaths: The 11 Rig Workers Who Died During the BP Deepwater Horizon Explosion," The Associated Press, November 15, 2012. http://www.huffingtonpost.com/2012/11/15/gulf-oil-spill-deaths_n_2139669.html.

165 Louisiana has been yielding marshland to the ocean at an extraordinary rate: "The Most Ambitious Environmental Lawsuit Ever," *New York Times Magazine,* October 14, 2014.

167 we sent dogs and chimps in unmanned spaceships as our surrogates: "The Horrible Thing That Happened to Enos the Chimp When He Orbited Earth 50 Years Ago," *The Atlantic*, November 29, 2011.

168 a drug used to treat streptococcal infections and labeled "Elixir Sulfanilamide" poisoned and killed more than one hundred patients: "Taste of Raspberries, Taste of Death, the 1937 Elixir Sulfanilamide Incident," U.S. Food and Drug Administration, Sulfanilamide Disaster, *FDA Consumer*, June 1981, http://www.fda.gov/AboutFDA/WhatWeDo/History/ProductRegulation/SulfanilamideDisaster/default.htm

168 Congress passed the Food, Drug, and Cosmetic Act of 1938: "Toxicity Testing in the 21st Century: A Vision and a Strategy," National Research Council. The National Academies Press, Washington, D.C., 2007.

168 Congress enacted the Toxic Substances Control Act: Report of the House Energy and Commerce Committee, Report 114–176, Washington, D.C., June 23, 2015.

168 Thalidomide underwent animal testing prior to its first test marketing on women: "The Return of Thalidomide," *BBC News—Inside Out*, January 31, 2014, http://www.bbc.co.uk/insideout/southwest/series7/thalidomide.shtml

168 Fialurdine . . . led to the death of five of fifteen human volunteers: Attarwala, H.

(2010). "TGN1412: From Discovery to Disaster," *Journal of Young Pharmacists*, 2(3), 332–336. http://doi.org/10.4103/0975–1483.66810.

168 all six human volunteers faced life-threatening conditions: Attarwala, H. (2010). "TGN1412: From Discovery to Disaster," *Journal of Young Pharmacists*, 2(3), 332–336. http://doi.org/10.4103/0975–1483.66810.

169 FDA reported that 92 percent of new drugs failed human safety standards: "Innovation or Stagnation: Challenge and Opportunity on the Critical Path to New Medical Products: Challenges and Opportunities Report—March 2004," US Food and Drug Administration, March 16, 2004, http://www.fda.gov/ScienceResearch/SpecialTopics/CriticalPathInitiative/CriticalPathOpportunitiesReports/ucm077262.htm.

170 Senator Vitter ominously described it as "a new oil spill every day": "Sen. David Vitter takes aim at BP, Coast Guard over Gulf oil spill cleanup," *New Orleans Times Picayune*, October 2, 2012.

171 Under the 1976 Toxic Substances Control Act (TSCA), the EPA has long been charged with preventing an "unreasonable risk of injury to health or the environment."

171 the EPA conducted comprehensive assessments of only two hundred chemicals: Report of the Senate Committee on Environment and Public Works, Frank R. Lautenberg Chemical Safety for the Twenty-First Century Act, Washington, D.C., June 18, 2015.

171 the EPA has an inventory of 85,000 chemicals: Report of the Senate Committee on Environment and Public Works, Frank R. Lautenberg Chemical Safety for the Twenty-First Century Act, Washington, D.C., June 18, 2015.

172 the use of nearly two million gallons of dispersants: "Half of oil from BP's Gulf of Mexico spill 'may still be on sea floor,' " *The Telegraph*, November 15, 2015.

172 Animal toxicity studies can take three to five years per chemical: "2-Year Study Protocol, Descriptions of NPT Study Types," National Toxicology Program, US Department of Health and Human Services.

172 A single evaluation of a potential carcinogen requires 400 rats and 400 mice: "2-Year Study Protocol, Descriptions of NPT Study Types," National Toxicology Program, US Department of Health and Human Services.

173 The EPA did the lab work and then published a report on eight dispersants in weeks: "Comparative Toxicity of Eight Oil Dispersant Products on Two Gulf of Mexico Aquatic Test Species," US Environmental Protection Agency, Office of Research and Development, National Health and Environmental Effects Research Laboratory, Michael J. Hemmer, Mace G. Barron, and Richard M. Greene, June 30, 2010.

174 The new approach would generate more relevant data: "Toxicity Testing in the 21st Century: A Vision and a Strategy," National Research Council. The National Academies Press, Washington, D.C., 2007.

174 modular in vitro human immune systems: "Lessons from Toxicology: Developing a 21st-Century Paradigm for Medical Research," *Environmental Health Perspectives*, 123(11), November 2015, .http://dx.doi.org/10.1289/ehp.1510345.

174 Canada was the innovator in using these tools: "Canada: moving forward on

chemicals assessment," Chemical Watch, May 2014. https://chemicalwatch.com/19746/canada-moving-forward-on-chemicals-assessment.

175 HSI collaborated with US and EU regulatory authorities and industry: "Highest EU Court To Interpret Cosmetic Animal-Test Ban's Scope," The Rose Sheet, Dec. 18, 2014.

176 the Senate approved a new standard . . . by a unanimous vote: Cana, Timothy, "Senate passes overhaul of chemical rules," *The Hill*, December 17, 2015.

176 won approval of legislation to create a government body, known as ICCVAM: H.R. 4281 (106th): ICCVAM Authorization Act of 2000, govtrack.us, https://www.govtrack.us/congress/bills/106/hr4281.

177 it may be justifiable to skip the animal model assessment of efficacy altogether: "Avoiding Animal Testing," *The Scientist*, December 1, 2011.

177 "We have moved away from studying human disease in humans" : "Ex-Director Zerhouni Surveys Value of NIH Research," LXV(13), http://nihrecord.od.nih.gov/newsletters/2013/06_21_2013/story1.htm

178 "turn observations in the laboratory, clinic and community into interventions: National Center for Advancing Translation Sciences, 2014 Report. https://ncats.nih.gov/files/NCATS_2014_report.pdf.

179 "One can lay a lot of the clinical failures at the paws of the animal data: "Abandoning Linear Thinking," *Biocentury Innovations*, August 27, 2015.

179 recently invested $70 million to develop up to ten organs on chips: "Wyss Institute to Receive up to $37 Million from DARPA to Integrate Multiple Organ-on-Chip Systems to Mimic the Whole Human Body," July 24, 2012, http://wyss.harvard.edu/viewpressrelease/91/wyss-institute-to-receive-up-to-37-million-from-darpa-to-integrate-multiple-organonchip-systems-to-mimic-the-whole-human-body.

181 COTY, which bought the beauty division of Procter & Gamble: "Humane Society Gears Up For Renewed Push Behind U.S. Animal-Test Ban," The Rose Sheet, August 19, 2015.

181 China lifted that requirement for cosmetics manufactured domestically: "U.S. Or China? Humane Cosmetics Act Could Force Multinationals To Choose," The Rose Sheet, Aug. 13, 2015.

182 L'Oréal made headlines in the spring of 2015 when the cosmetic giant: Coleman, Kelly P.; McNamara, Lori R.; Grailer, Thomas P.; Willoughby, Jamin A. Sr.; Keller, Donald J.; Patel, Prakash; Thomas, Simon; and Dilworth, Clive. *Applied In Vitro Toxicology*. June 2015, 1(2): 118–130, doi:10.1089/aivt.2015.0007. L'Oréal played a pioneering role in developing alternatives to animal testing, L'Oreal, http://www.lorealpredictive.com/en/mondial-network/pioneering.

182 $5.6 billion spent on the Project (in 2010 dollars) led to $796 billion: Tripp, Simon and Gruber, Martin, "Economic Impact of the Human Genome Project: How a $3.8 billion investment drove $796 billion in economic impact, created 310,000 jobs, and launched the genomic revolution," Battelle Memorial Institute Technology Partnership Practice, May 2011.

182 The global pharmaceutical industry generates net income of $980 billion: "In Vitro Toxicity Testing Market to Exhibit 15.3% CAGR to 2018 as Animal Test-

ing is Phased Out." See more at: http://www.industrytoday.co.uk/market-research-industry-today/in-vitro-toxicity-testing-market-to-exhibit-153-cagr-to-2018-as-animal-testing-is-phased-out/42449#sthash.RvHdsACT.dpuf.

SIX: The Visible Hand and the Free Market: Humane Wildlife Management in the United States

189 In 2013, the National Park System received over 273 million visitors at 401 land areas: National Park Service, "Annual Visitation Summary Report for 2013," n.d., https://irma.nps.gov/Stats/SSRSReports/System Wide Reports/Annual Visitation Summary Report (1979—Last Calendar Year)?RptYear=2013&Fiscal=False.

190 those visitors spent $14.6 billion in surrounding gateway regions: Thomas, Catherine, Christopher Huber, and Lynne Koontz, "Executive Summary," *2013 National Park Visitor Spending Effects: Economic Contributions to Local Communities, States, and the Nation*, July 2014, http://www.nature.nps.gov/publications/nrpm.

190 "The broader value of 'ecosystems services' . . .": Diamandis, P. H., and Kotler, S. *Abundance: The Future Is Better than You Think*. New York: Free Press. 2012; "The Value of the World's Ecosystem Services and Natural Capital," *Nature*, 387, May 15, 1997, pp. 253–260.

190 the boosters for the proposed Maine Woods National Park: Miller, Kevin, "A national park or a national monument? North Woods groups shift focus," *Portland Press Herald*, November 29, 2015.

191 "visitors to Korea's forests increased from 9.4 million in 2010 to 12.8 in 2013": Williams, Florence, "This Is Your Brain on Nature, National Geographic, Vol. 229, No. 1, January 2016, p. 62.

191 The Harvard biologist E.O. Wilson calls it "biophilia": Edward O. Wilson, *Biophilia: The Human Bond with Other Species* (Cambridge, MA: The President and Fellows of Harvard College, 1984), 1.

192 During his nearly eight years in office, he created: National Park Service, "Theodore Roosevelt and Conservation," n.d., http://www.nps.gov/thro/learn/historyculture/theodore-roosevelt-and-conservation.htm.

192 Altogether, Theodore Roosevelt placed 230 million acres of land under permanent public protection: National Park Service, "Theodore Roosevelt and Conservation," n.d., http://www.nps.gov/thro/learn/historyculture/theodore-roosevelt-and-conservation.htm.

193 In 2013, Yellowstone Park alone received 3.2 million visits: Thomas, Catherine, Christopher Huber, and Lynne Koontz, "Table 3. Visits, spending and economic contributions to local economies of NPS visitor spending," *2013 National Park Visitor Spending Effects: Economic Contributions to Local Communities, States, and the Nation*, July 2014, http://www.nature.nps.gov/publications/nrpm.

193 Yosemite received 3.7 million visits: Thomas, Catherine, Christopher Huber, and Lynne Koontz, "Table 3. Visits, spending and economic contributions to local economies of NPS visitor spending," *2013 National Park Visitor Spending Effects: Economic Contributions to Local Communities, States, and the Nation*, July 2014, http://www.nature.nps.gov/publications/nrpm.

193 Great Smoky Mountain Park in eastern Tennessee: Thomas, Catherine, Christopher Huber, and Lynne Koontz, "Table 3. Visits, spending and economic contributions to local economies of NPS visitor spending," *2013 National Park Visitor Spending Effects: Economic Contributions to Local Communities, States, and the Nation*, July 2014, http://www.nature.nps.gov/publications/nrpm.

193 By comparison, the fifteen-to-twenty-thousand people who go to Isle Royale: Thomas, Catherine, Christopher Huber, and Lynne Koontz, "Table 3. Visits, spending and economic contributions to local economies of NPS visitor spending," *2013 National Park Visitor Spending Effects: Economic Contributions to Local Communities, States, and the Nation*, July 2014, http://www.nature.nps.gov/publications/nrpm.

194 "three remaining wolves may struggle to reproduce . . .": Peters, Gary, Letter to Jonathan B. Jarvis, Director of National Park Service. May 29, 2015, http://www.peters.senate.gov/sites/default/files/150529_Letter_Wolves_IsleRoyale.pdf.

194 "Unless the NPS acts quickly . . .": "Peters Urges Action to Address Declining Wolf Population at Isle Royale National Park," May 29, 2015, http://www.peters.senate.gov/newsroom/press-releases/peters-urges-action-to-address-declining-wolf-population-at-isle-royale-national-park.

194 they rejected two statewide referendums: Matheny, Keith, "Voters Have Their Say on Wolf Hunt, but Will it Matter?," *Detroit Free Press*, November 5, 2014, http://www.freep.com/story/news/local/michigan/2014/11/04/michigan-wolf-hunt/18504391/.

195 . . . there were more than one hundred packs surviving there: Michigan Wolf Management Plan (Updated 2015), Michigan Department of Natural Resources, June 11, 2015.

195 With 300,000 people in the Upper Peninsula and just more than 600 wolves: Lake Superior Community Partnership, "Demographics & Economic Indicators," n.d., http://marquette.org/demographics-economic-indicators/.

195 They trotted out the experience of an Upper Peninsula farmer: Barnes, John, "John Koski, Part 1: Tour the farm with more wolf attacks than anyone in Michigan's Upper Peninsula," *MLive*, November 4, 2013, ww.mlive.com/news/index.ssf/2013/11/john_koski_part_1_tour_the_far.html#incart_river_default.

195 Koski's farm was the site of more than 60 percent of all wolf attacks: Barnes, John, "John Koski, Part 1: Tour the farm with more wolf attacks than anyone in Michigan's Upper Peninsula," *MLive*, November 4, 2013, ww.mlive.com/news/index.ssf/2013/11/john_koski_part_1_tour_the_far.html#incart_river_default.

196 . . . according to a months-long investigation: Barnes, John, "Michigan's Wolf Hunt: How Half-Truths, Falsehoods and One Farmer Distorted Reasons for Historic Hunt," *MLive*, November 3, 2013, http://www.mlive.com/news/index.ssf/2013/11/michigans_wolf_hunt_how_half_t.html.

196 The state even gave him some guard donkeys: Barnes, John, "Michigan's Wolf Hunt: How Half-Truths, Falsehoods and One Farmer Distorted Reasons for Historic Hunt," *MLive*, December 9, 2013, http://www.mlive.com/news/index.ssf/2013/11/michigans_wolf_hunt_how_half_t.html.

196 The reporter found that the Michigan Department of Natural Resources:

Oosting, Jonathan, "Michigan Wolf Hunting: Everything You Need to Know about Proposal 1 and 2 on the 2014 Ballot," *MLive*, October, 31, 2004, http://www.mlive.com/lansing-news/index.ssf/2014/10/michigan_2014_election_everyth_2.html.

196 "government half-truths, falsehoods, and livestock numbers skewed by a single farmer . . .": Barnes, John, "Michigan's Wolf Hunt: How Half-Truths, Falsehoods and One Farmer Distorted Reasons for Historic Hunt," *MLive*, November 3, 2013, http://www.mlive.com/news/index.ssf/2013/11/michigans_wolf_hunt_how_half_t.html.

196 Koski pled no contest to charges of animal neglect: Barnes, John, "No Jail for Wolf Figure, Almost $1,900 in Fines Ordered against Farmer Accused of Leaving Cattle Vulnerable," *MLive*, May 9, 2014, http://www.mlive.com/news/index.ssf/2014/05/no_jail_for_wolf_figure_almost.html.

196 a US District Court judge ruled . . . that the federal government's action: Oosting, Jonathan, "Delist or Downlist? Michigan Wolf Debate Rages on Following Federal Ruling that Blocked Hunting," *MLive*, January 28, 2015, http://www.mlive.com/lansing-news/index.ssf/2015/01/delist_or_downlist_michigan_wo.html.

196 The court restored federal protection: Karnowski, Steve, "Bill Would Remove Federal Protections for Wolves in Four States, Including Minnesota," *Star Tribune*, January 13, 2015, http://www.startribune.com/bill-would-remove-federal-protections-for-wolves-in-four-states-including-minnesota/288446441/.

196 Republican Dan Benishek introduced legislation to overturn the court ruling: Office of Congressman Dan Benishek, "Dr. Benishek Introduces Bill to Address Gray Wolves," February 12, 2015, http://benishek.house.gov/press-release/dr-benishek-introduces-bill-address-gray-wolves.

197 Casperson got in trouble for exaggerating the case against wolves: Oosting, Jonathan, "Michigan Senator Apologizes for Fictional Wolf Story in Resolution: 'I am Accountable, and I am Sorry,' " *MLive*, November 7, 2013, http://www.mlive.com/news/index.ssf/2013/11/michigan_senator_apologizes_fo.html.

197 He concocted a menacing story out of a rather benign incident: Oosting, Jonathan, "Michigan Senator Apologizes for Fictional Wolf Story in Resolution: 'I am Accountable, and I am Sorry,' " *MLive*, November 7, 2013, http://www.mlive.com/news/index.ssf/2013/11/michigan_senator_apologizes_fo.html.

197 Casperson highlighted this story in introducing a resolution: Michigan Legislature, "Senate Resolution 0039 (2011)," n.d., http://legislature.mi.gov/doc.aspx?2011-SR-0039.

197 "there were no children in the backyard that day . . .": Barnes, John, "The Michigan Myth: How Lawmakers Turned this True Wolf Story into Fiction," *MLive*, November 3, 2013, http://www.mlive.com/news/index.ssf/2013/11/the_michigan_myth_read_the_tru.html.

197 Tom Casperson took enough heat over his statement: Oosting, Jonathan, "Michigan Senator Apologizes for Fictional Wolf Story in Resolution: 'I am Accountable, and I am Sorry,' " *MLive*, November 7, 2013, http://www.mlive.com/news/index.ssf/2013/11/michigan_senator_apologizes_fo.html.

197 "We are concerned for our children's safety . . .": Neher, Jake, "State Senate

Urges Congress to Remove Gray Wolves from Endangered List," WKAR, February 10, 2015, http://www.tinyurl.com/q6mztk5.

197 of America's big predators, wolves have proved among the least menacing : Viegas, Jennifer, Wolf Attacks More Myth Than Reality, Discovery.com, http://news.discovery.com/animals/endangered-species/wolf-attacks-more-myth-than-reality-150311.htm, and Kullgren, Ian, "Department of Fish and Wildlife says there have been no wolf-related deaths in the Rockies," Politifact.com, December 16, 2011. http://www.politifact.com/oregon/statements/2011/dec/16/oregon-department-fish-and-wildlife/department-fish-and-wildlife-says-there-have-been-/.

198 "Dogs absolutely turned the tables": "Dogs Decoded: How smart are dogs, and what makes them such ideal companions?" PBS Documentary, July 3, 2013. http://www.pbs.org/wgbh/nova/nature/dogs-decoded.html.

199 "It's hard to see how early herders would have moved . . .": Rowley-Conwy, Peter, "Dogs Decoded," *NOVA*, November 9, 2010, http://www.pbs.org/wgbh/nova/nature/dogs-decoded.html.

199 "the value of their [wolves and coyotes] hides . . .": Hampton, Bruce, *The Great American Wolf* (New York: Henry Holt and Company, Inc., 1997), 113.

200 they are responsible for a very small amount of killing: : Keefover, Wendy, "Government Report: Less Than 1% of Cattle Killed by Native Carnivores and Domestic Dogs," Huffington Post, May 18, 2011. http://www.huffingtonpost.com/wendy-keefoverring/native-carnivore-controls-unnecessary_b_863717.html "Cattle Death Loss Inventory," National Agricultural Statistics Service, 2010. http://usda.mannlib.cornell.edu/usda/current/CattDeath/CattDeath-05–12–2011.pdf.

200 A 2014 Washington State University study: Wielgus, Robert, and Kaylie Peebles, "Effects of Wolf Mortality on Livestock Depredations," *PLoS ONE* (2014), 9.

200 there are roughly 1.2 million deer–vehicle collisions: State Farm, "Top Five States For Deer-Related Collisions Named," n.d., https://www.statefarm.com/retirees/news/top-states-for-deer-collisions; State Farm, "When Bumpers Meet Antlers," n.d., http://teendriving.statefarm.com/learning-to-drive/driving-with-a-permit/when-bumpers-meet-antlers.

201 Michigan typically accounts for about fifty thousand of these collisions a year: Michigan State Police, "Vehicle-Deer Crashes," n.d., http://www.michigan.gov/msp/0,,7–123–72297_64773_22760–95455—,00.html; Michigan Traffic Craft Facts, "Upper Peninsular Monthly and Seasonal Rates for Motor Vehicle-Deer Crashes," 2014, http://publications.michigantrafficcrashfacts.org/Upper_Peninsula/2014/UP-deer+3.pdf.

201 "Wolves provide a firewall against new diseases in deer . . .": Forgrave, Will, "Western High School Graduate Working toward Statewide Referendum on Wolf Hunting Bill," *MLive*, February 27, 2013, http://www.mlive.com/news/jackson/index.ssf/2013/02/western_high_school_grad_worki.html.

201 there are more than four hundred deer farms in the state: Wisconsin Department of Natural Resources, "White-Tailed Deer Farming," 2015, http://dnr.wi.gov/topic/wildlifehabitat/regulations.html.

201 "So far CWD has not spread into areas inhabited by wolves . . .": Forgrave, Will,

"Western High School Graduate Working toward Statewide Referendum on Wolf Hunting Bill," *MLive*, February 27, 2013, http://www.mlive.com/news/jackson/index.ssf/2013/02/western_high_school_grad_worki.html.

201 Each year, thousands of wildlife watchers . . . bringing in $35 million to the Yellowstone region annually: Duffield, John, Chris Neher, and David Patterson, "Wolves and People in Yellowstone: Impacts on the Regional Economy," September 2006, 51, http://www.defenders.org/publications/wolves_and_people_in_yellowstone.pdf.

202 the International Wolf Center in Ely, Minnesota brings $3 million each year: Schaller, David, "Section 6: Impact of the International Wolf Center on Tourism and the Economy," *The Ecocenter as a Tourist Attraction: Ely and the International Wolf Center*, http://www.wolf.org/wow/united-states/minnesota/attitudes-and-issues-6/.

202 "I think that there are several areas where wolves are already playing a role . . .": Peterson, Rolf, "Testimony re. Senate Bill 1350," November 6, 2012, http://legislature.mi.gov/documents/2011–2012/CommitteeDocuments/Senate/Natural%20Resources%20Environment%20and%20Great%20Lakes/Testimony/2012-SCT-NAT_-11–08–1-24-Rolf%20Peterson%20%20(Con).PDF.

203 only five thousand or so wolves survive in the lower 48 states: US Fish & Wildlife Service, "Gray Wolf (*Canis lupus*) Current Population in the United States," 2015, http://www.fws.gov/midwest/wolf/aboutwolves/WolfPopUS.htm.

203 trappers and trophy hunters . . . killed off seventeen family units: Wisconsin Department of Natural Resources. 2014. "Green Sheet"—Request approval of a wolf harvest quota and establish the number of licenses to issue for the 2014–2015 wolf hunting and trapping season. Natural Resources Board, Cathy Stepp, Secretary.

204 In 2015, dozens of world-renowned wildlife biologists and scientists wrote to Congress: Ahern, Louise, "Who Should Manage Michigan's Wolves?", *Detroit Free Press*, February 19, 2015, http://on.freep.com/1EtxoBY.

204 "In recognition of the ecological benefits wolves bring . . .": Ahern, Louise, "Who Should Manage Michigan's Wolves?", *Detroit Free Press*, February 19, 2015, http://on.freep.com/1EtxoBY.

205 In the Sand Wash Basin, the BLM gives three grazing allotments: Personal communications, Jerome Fox, August 2015.

206 rounding up sixty-two mares and hand-injecting them: Bureau of Land Management, "Wild Horse Management Activities Continue into Fall at Sand Wash," June 16, 2010, http://www.blm.gov/co/st/en/BLM_Information/newsroom/2010/wild_horse_management.html.

207 Prior to 2013, PZP was classed as an experimental drug: Fox, Jerome, "Describe the Proposed Action," *Documentation of Land Use Plan Conformance and NEPA Adequacy,* 2013, 2, http://www.blm.gov/style/medialib/blm/co/information/nepa/little_snake_field/2013_documents.Par.33714.File.dat/DOI-BLM-CO-N010–2013–0029-DNA.pdf.

207 now it is registered by the Environmental Protection Agency under the brand name ZonaStat-H: Cima, Greg, "Vaccine Could Reduce Wild Horse Overpopula-

tion," *JAVMA News*, 2012, https://www.avma.org/News/JAVMANews/Pages/120415k.aspx.

209 According to the BLM, there are about sixty thousand free-roaming wild horses and burros in the West: Bureau of Land Management, "Wild Horse and Burro Quick Facts," n.d., http://www.blm.gov/wo/st/en/prog/whbprogram/history_and_facts/quick_facts.html.

209 The federal Wild Horse and Free-Roaming Burro Act of 1971 forbids: The Wild Free-Roaming Horses and Burros Act of 1971 (Public Law 92–195), http://www.wildhorseandburro.blm.gov/92–195.htm.

209 This law was enacted after a campaign by Nevadan Velma Johnston: Barendse, Michael, "Johnston, Velma B.," Learning to Give, n.d., http://www.learningtogive.org/resources/johnston-velma-b.

209 Johnston started this historic campaign after one morning on her way to work in Reno: Barendse, Michael, "Johnston, Velma B.," Learning to Give, n.d., http://www.learningtogive.org/resources/johnston-velma-b.

210 In the forty-five years since the law was enacted: Bureau of Land Management, "Wild Horse and Burro Quick Facts," n.d., http://www.blm.gov/wo/st/en/prog/whbprogram/history_and_facts/quick_facts.html.

210 The BLM, in the late years of the George W. Bush administration: The Bureau of Land Management, Wild Horse and Burro Program Data. http://www.blm.gov/wo/st/en/prog/whbprogram/herd_management/Data.html.

211 By 2015, the number of captive wild horses had reached 47,000: Bureau of Land Management, Wild Horse and Burro Quick Facts, March 1, 2015, http://www.blm.gov/wo/st/en/prog/whbprogram/history_and_facts/quick_facts.html.

211 the agency was now responsible for more than 100,000 horses: Bureau of Land Management, Wild Horse and Burro Quick Facts, March 1, 2015, http://www.blm.gov/wo/st/en/prog/whbprogram/history_and_facts/quick_facts.html.

211 "Regularly removing horses holds population levels below food-limiting carry capacity . . .": Committee to Review the Bureau of Land Management Wild Horse and Burro Management Program, Summary, "Using Science to Improve the BLM Wild Horse and Burro Program: A Way Forward," 2013, 5, http://www.fs.usda.gov/Internet/FSE_DOCUMENTS/stelprd3796106.pdf.

211 "It is clear that the status quo of continually removing free-ranging horses . . .": Committee to Review the Bureau of Land Management Wild Horse and Burro Management Program, Preface, "Using Science to Improve the BLM Wild Horse and Burro Program: A Way Forward," 2013, vii, http://www.fs.usda.gov/Internet/FSE_DOCUMENTS/stelprd3796106.pdf.

211 That cost is nearly $1,000 per horse per year: Bureau of Land Management, Wild Horse and Burro Quick Facts, March 1, 2015, http://www.blm.gov/wo/st/en/prog/whbprogram/history_and_facts/quick_facts.html.

212 "Investing in science-based management approaches would not solve the problem instantly . . .": Committee to Review the Bureau of Land Management Wild Horse and Burro Management Program, Summary, "Using Science to Improve the BLM Wild Horse and Burro Program: A Way Forward," 2013, 12, http://www.fs.usda.gov/Internet/FSE_DOCUMENTS/stelprd3796106.pdf.

212 "The porcine zona pellucida (PZP) vaccine . . .": Committee to Review the Bureau of Land Management Wild Horse and Burro Management Program, Fertility Management, "Using Science to Improve the BLM Wild Horse and Burro Program: A Way Forward," 2013, 113, http://www.fs.usda.gov/Internet/FSE_DOCUMENTS/stelprd3796106.pdf.

214 For every wild horse . . . there are sixty-seven cattle and sheep: Center for Biological Diversity Study: Livestock Grazing on Public Lands Cost Taxpayers $1 Billion Over Past Decade, January 28, 2015. http://www.biologicaldiversity.org/news/press_releases/2015/grazing-01–28–2015.html

215 Wildlife Services has been amassing a body count of 100,000 coyotes a year: United States Department of Agriculture, "Wildlife Damage," n.d., https://www.aphis.usda.gov/wps/portal/aphis/ourfocus/wildlifedamage.

215 Taxpayers spend more than $60 million a year: Animal Plant Health Inspection Service, Wildlife Damage Management, https://www.aphis.usda.gov/wildlife_damage/budget_info.shtml.

SEVEN: **Global Growth Stocks: Elephants, Lions, Great Apes, Whales, Sharks, and Other Living Capital**

223 Kenya alone has fifty-one terrestrial national parks and reserves: Kenya Wildlife Service, "Overview," http://www.kws.go.ke/content/overview-0/.

223 worth $48 million a year in gate fees and billions more in related commerce: Wanyonyi, Edwin, "Mobilizing Resources for Wildlife Conservation in Kenya beyond the 21st Century," *The George Wright Forum*, 2012, 29, http://www.georgewright.org/291wanyonyi.pdf

223 Wildlife tourism yields . . . 25 percent Kenya's GDP: Akama, John, Shem Maingi, and Blanca Camargo, "Wildlife Conservation, Safari Tourism and the Role of Tourism Certification in Kenya: A Postcolonial Critique," *Tourism Recreation Research*, 51 (2011), http://trrworld.org/pdfs/96k18smz3543t18cr12bts3642o15fq.pdf.

223 more than 300,000 jobs: World Bank, "Kenya's Tourism: Polishing the Jewel," 2010, http://siteresources.worldbank.org/KENYAEXTN/Resources/Tourism_Report-ESW_Kenya_Final_May_2010.pdf

223 Kenya has forbidden sport hunting since 1977: Honey, Martha, *Ecotourism and Sustainable Development, Second Edition: Who Owns Paradise*, Island Press, 2008. p. 297. "

223 Walter Palmer, an American dentist working with two professional guides: Dorian, Marc, Lauren Putrino, and Cat Rakowski, "What Happened in the Harrowing Hours Before Cecil the Lion Was Killed," *ABC News*, August 13, 2015, http://abcnews.go.com/International/happened-harrowing-hours-cecil-lion-killed/story?id=33044279.

223 Palmer let loose with a broadhead arrow: Walsh, Paul, "Full Transcript: Walter Palmer Speaks About Cecil the Lion Controversy," *Minneapolis Star Tribune*, September 7, 2015, http://strib.mn/1VJipNp.

224 Palmer and his two guides chose not to give chase: Dorian, Marc, Lauren Putrino, and Cat Rakowski, "What Happened in the Harrowing Hours Before

Cecil the Lion was Killed," *ABC News*, August 13, 2015, http://abcnews.go.com/International/happened-harrowing-hours-cecil-lion-killed/story?id=33044279.

224 The lion . . . suffered for at least a dozen hours: Walsh, Paul, "Full Transcript: Walter Palmer Speaks About Cecil the Lion Controversy," *Minneapolis Star Tribune,* September 7, 2015, http://strib.mn/1VJipNp.

224 they realized Cecil was wearing a radio collar: Walsh, Paul, "Full Transcript: Walter Palmer Speaks About Cecil the Lion Controversy," *Minneapolis Star Tribune*, September 7, 2015, http://strib.mn/1VJipNp.

224 Instead of reporting the incident, the threesome tried to cover up the killing: Kassam, Ashifa, and Jessica Glenza, "Killer of Cecil the Lion was Dentist from Minnesota, Claim Zimbabwe Officials," *The Guardian*, July 28, 2015, http://gu.com/p/4b3h8/sbl.

224 This wasn't his first alleged attempt to cover up a wildlife crime. He had a felony guilty plea on his record, after illegally slaying an enormous trophy black bear: Smith, Alexander, and Federico-O'Murchu, Sean, "Cecil the Lion Killer Walter James Palmer Has Bear-Related Felony Record," *NBC News*, July 29, 2015, http://www.nbcnews.com/news/world/cecil-lion-killer-walter-james-palmer-has-bear-related-felony-n400226.

224 then attempting to bribe his guides to lie about the location of the kill: Bucktin, Christopher, "Cecil the Lion: Dentist Walter Palmer Tried to BRIBE Guides to 'Cover Up' His Illegal Bear Hunt," *Mirror*, August 14, 2015, http://www.mirror.co.uk/news/world-news/cecil-lion-dentist-walter-palmer-6255227.

224 "This trophy hunting is destroying our wildlife, and for what really?": Ndebele, Hazel, "Bigwigs' Rich Pickings from Game Hunting," *Zimbabwe Independent*, July 31, 2015, http://www.theindependent.co.zw/2015/07/31/bigwigs-rich-pickings-from-game-hunting/.

224 A 2013 study revealed that trophy hunting in 2011 generated just 0.2 percent of the nation's GDP: Campbell, Roderick, "The $200 Million Question: How Much Does Trophy Hunting Really Contribute to African Communities?", *Economists at Large*, 2013, 11–12, http://www.hsi.org/assets/pdfs/200-million-question-ecolarge-trophy-hunting-study.pdf.

224 "We have lost a lion which marketed our country . . . Cecil was killed for only US$50,000 . . .": Ndebele, Hazel, "Bigwigs' Rich Pickings from Game Hunting," *Zimbabwe Independent*, July 31, 2015, http://www.theindependent.co.zw/2015/07/31/bigwigs-rich-pickings-from-game-hunting/.

225 Americans like Walter Palmer kill more than 700 African lions and about 600 elephants every year: Bedard, Paul, Humane Society: Obama has effectively ended 'canned' African lion hunts, Americans kill 90%; *The Washington Examiner*, December 30, 2015. And Telecky, Teresa, "Hunting is a Setback to Wildlife Conservation," *Earth Island Journal*, Summer 2014, http://www.earth-island.org/journal/index.php/eij/article/hunting_is_a_setback_to_wildlife_conservation/

225 Palmer's guide said that the Minnesota-based hunter also wanted to kill a trophy elephant on that trip: "Cecil the lion's killer Walter Palmer wanted to stalk an elephant next—but couldn't find one big enough,' " *The Telegraph*, July 31, 2015, http://www.telegraph.co.uk/news/worldnews/africaandindianocean/

zimbabwe/11773653/Cecil-the-lions-killer-Walter-Palmer-wanted-to-stalk-an-elephant-next-but-couldnt-find-one-big-enough.html.

225 the United States suspended any imports of sport-hunted elephant tusks: US Fish & Wildlife Service, "Service Suspends Import of Elephant Trophies from Tanzania and Zimbabwe," Press Release, April 4, 2014, http://www.fws.gov/news/ShowNews.cfm?ID=2E6FF2A2-E10F-82BC-DAE08807810E3C6B

225 Zimbabwe restricted for a few weeks the killing: Mutsake, Farai, "Zimbabwe suspends hunting, investigates other lion killing in area where Cecil the lion killed," *Minneapolis Star Tribune*, August 1, 2015, http://www.startribune.com/zimbabwean-authorities-restrict-hunting-after-lion-killing/320381511/.

225 Three months later, a German trophy hunter came to Zimbabwe: "Biggest elephant killed in Africa for almost 30 years brings back memories of Cecil the lion," *The Telegraph*, October 16, 2015, http://www.telegraph.co.uk/news/worldnews/africaandindianocean/zimbabwe/11934535/Huge-tusked-African-elephant-killed-by-german-hunter-in-Zimbabwe.html.

225 the tyrant threw himself an extravagant birthday party: "Robert Mugabe eats a zoo for " 'obscene' 91st birthday party," *The Independent*, March 2, 2015. http://www.independent.co.uk/news/people/robert-mugabe-eats-a-zoo-for-obscene-91st-birthday-party-10077805.html.

226 Safari Club International, with its twenty-nine "hunting achievement" and dozens of "Inner Circle" awards: Safari Club International, "Awards 2015," *Safari Magazine*, 2015, http://member.scifirstforhunters.org/static/WHA/15/.

226 a trophy hunter must kill as many as 322 different species and subspecies: Safari Club International, "Awards 2015," *Safari Magazine*, 2015, http://member.scifirstforhunters.org/static/WHA/15/.

226 the most coveted of the awards is the "Africa Big Five,": Safari Club International, "Awards 2015," *Safari Magazine*, 2015, http://member.scifirstforhunters.org/static/WHA/15/.

227 Lufthansa KLM, Singapore Airlines, Emirates, and Virgin Atlantic already had policies in place: du Venage, Gavin, "US Airlines Follow Etihad, Emirates in Banning Big-Game Trophy Cargo", *The National*, August 11, 2015, http://www.thenational.ae/business/aviation/us-airlines-follow-etihad-emirates-in-banning-big-game-trophy-cargo.

227 Delta announced it "will officially ban . . .": Staff Writer, "Shipments of Lion, Leopard, Elephant, Rhino, Buffalo Trophies Banned," News Hub, August 3, 2015, http://news.delta.com/shipments-lion-leopard-elephant-rhino-buffalo-trophies-banned.

227 American, United, and Air Canada followed suit: du Venage, Gavin, "US Airlines Follow Etihad, Emirates in Banning Big-Game Trophy Cargo," *The National*, August 11, 2015, http://www.thenational.ae/business/aviation/us-airlines-follow-etihad-emirates-in-banning-big-game-trophy-cargo.

227 DHL was next: Martin, Hugo, "After Cecil, Some Air Freight, Carriers Still Accept Animal Trophies," *Los Angeles Times*, August 9, 2015, http://www.latimes.com/business/la-fi-after-cecil-some-air-freight-carriers-still-accept-animal-trophies-20150807-story.html.

227 "Trophy hunting feels like a relic of a bygone era . . .": Branson, Richard, "Big

Game is Worth More Alive Than Dead," *Virgin Airlines*, August 7, 2015, http://www.virgin.com/richard-branson/big-game-is-worth-more-alive-than-dead.

228 more than forty airlines and freight carriers got on board: "More than 40 airlines adopt wildlife trophy bans," *Sunday Times*, August 28, 2015.

228 France banned any imports of lion trophies: Vaughan, Adam, "France bans imports of lion hunt trophies," *The Guardian*, November 19, 2015.

228 US Fish and Wildlife Service, acting on a petition from HSUS: Department of the Interior, Fish and Wildlife Service, 50 CFR Part 17, Docket No. FWS-R9-ES-2012–0025; 450 003 0115, RIN 1018-BA29. https://www.gpo.gov/fdsys/pkg/FR-2015–12–23/html/2015–31958.htm

229 with corrupt management of lion-hunting programs in Tanzania: Goode, Erica, "A Lion Expert Who Isn't Inclined to Turn Tail," *New York Times*, December 28, 2015.

229 According to a 2013 report by the United Nations World Tourism Association: "Towards Measuring the Economic Value of the Wildlife Watching Tourism in Africa," World Tourism Association, Madrid, Spain, 2015. http://dtxtq4w60xqpw.cloudfront.net/sites/all/files/docpdf/unwtowildlifepaper.pdf.

229 trophy hunting receipts in Zimbabwe declined by 30 percent in 2015: Kudzai Kuwaza, Kudzai, "Zimbabwe: Safari Operations Decline By 30 Percent," AllAfrica, December 18, 2015.

230 American hunters can now only import African elephant trophies from Namibia and South Africa: US Fish & Wildlife Service, "Import of Sport-Hunted Trophies," Sport-Hunted Trophies, n.d., http://www.fws.gov/international/permits/by-activity/sport-hunted-trophies.html.

230 "if we do not take care of our animals . . .": Michler, Ian, "Botswana Kills Trophy Hunting—Ian Michler Reflects," *Africa Geographic*, November 13, 2012, http://africageographic.com/blog/botswana-kills-trophy-hunting-ian-michler-reflects/.

230 "Hunters only employ people during hunting season . . ." ; Jammot, Julie, "Botswana's Marauding Elephants Trigger Hunting Ban Debate," *Business Insider*, May 4, 2015, http://www.businessinsider.com/afp-botswanas-marauding-elephants-trigger-hunting-ban-debate-2015–5.

230 "Experience true ecotourism in a visionary country that outlaws hunting . . .": "Destination: Botswana," Adventure Women, n.d., http://www.adventurewomen.com/adventure-vacation/botswana-safari-okavango-delta/.

231 In 2002, then-President Omar Bongo created a new national park system: Mayell, Hillary, "Gabon to Create Huge Park System for Wildlife," *National Geographic News*, September 4, 2002, http://news.nationalgeographic.com/news/2002/09/0904_020904_gabonparks.html.

233 "A single dead elephant's tusks are estimated to have a raw value of $21,000 . . .": Hughes, Clarissa, "Rebranding Environmental Crime as a Crime Against Humanity," *Africa Geographic*, October 23, 2014, http://africageographic.com/blog/rebranding-environmental-crime-as-a-crime-against-humanity/.

234 "60 percent of diseases that afflict humans start in animals," Laurance, Jeremy, "Deadly Animal Diseases Poised to Infect Humans," *The Independent*, January 4, 2010, http://www.independent.co.uk/news/science/deadly-animal-diseases-poised-to-infect-humans-1856777.html.

234 "don't touch your friends . . .": Public Radio International, "This Catchy West African Dance Tune Carries a Public Health Message About Ebola," *The World*, May 28, 2014, http://www.pri.org/stories/2014-05-28/catchy-west-african-dance-tune-carries-public-health-message-about-ebola.

234 it's critical to win "the hearts and minds" of local communities: Tsavo Trust, newsletter, June 2014, http://issuu.com/tsavotrust/docs/140709_tsavo_trust_newsletter_v8.

235 "they think we're the enemy": Scully, Matthew, "Inside the Global Industry That's Slaughtering Africa's Elephants," *The Atlantic*, June 6, 2013, http://www.theatlantic.com/international/archive/2013/06/inside-the-global-industry-thats-slaughtering-africas-elephants/276582/.

236 Kenya had an estimated 167,000 elephants in 1979: Nzwili, Fredrick, "Kenya's Elephants May Vanish in 10 Years, Warns Prominent Naturalist," *The Christian Science Monitor*, July 27, 2013, http://www.csmonitor.com/World/Africa/2013/0727/Kenya-s-elephants-may-vanish-in-10-years-warns-prominent-naturalist.

236 There were roughly 200,000 elephants in Angola decades ago: "Status of Elephant Populations, Levels of Illegal Killing and the Trade in Ivory: A Report to the Standing Committee of CITES," CITES, SC61 Doc. 44.2 (Rev. 1) Annex 1, https://cites.org/eng/com/sc/61/E61-44-02-A1.pdf; "Elephants in the Dust: The African Elephant Crisis, CITES, A Rapid Response Assessment. United Nations Environment Programme, GRID-Arendal. Norway, 2013. https://cites.org/common/resources/pub/Elephants_in_the_dust.pdf.

237 Today, fewer than 400,000 remain: International Elephant Foundation, "African Elephants," n.d., http://www.elephantconservation.org/elephants/african-elephants/.

237 perhaps 100,000 elephants killed a year during the worst of it: World Wildlife Fund, "Threats to African Elephants," n.d., http://wwf.panda.org/what_we_do/endangered_species/elephants/african_elephants/afelephants_threats/.

237 "more than 85 percent of the forest elephant ivory seized between 2006 and 2014 . . .": Feltman, Rachel, "Scientists Have Used DNA Tests to Track Africa's Worst Elephant Poaching Spots," *The Washington Post*, June 18, 2015, https://www.washingtonpost.com/news/speaking-of-science/wp/2015/06/18/scientists-have-used-dna-tests-to-track-africas-worst-elephant-poaching-spots/.

238 "more than 85 percent of the savanna elephant ivory . . .": Feltman, Rachel, "Scientists Have Used DNA Tests to Track Africa's Worst Elephant Poaching Spots," *The Washington Post*, June 18, 2015, https://www.washingtonpost.com/news/speaking-of-science/wp/2015/06/18/scientists-have-used-dna-tests-to-track-africas-worst-elephant-poaching-spots/.

238 Ranking 185th out of 187 countries: United Nations, "Mozambique," *Human Development Report 2013*, http://hdr.undp.org/sites/default/files/Country-Profiles/MOZ.pdf.

238 estimated 9,345 elephants were poached out of a population: "Mozambique," *Human Development Report 2013*, http://hdr.undp.org/sites/default/files/Country-Profiles/MOZ.pdf.

238 Tanzania, which once had the largest domestic elephant population: Environ-

mental Investigation Agency, "Crime & Corruption Behind Tanzania's Elephant Meltdown," November 6, 2014, https://eia-international.org/crime-corruption-behind-tanzanias-elephant-meltdown.

238 the elephant population has fallen from 109,000 in 1976 to 13,084 in 2013: Laing, Aislinn, "Tanzania's Elephant Catastrophe: 'We recalculated about 1,000 times because we didn't believe what we were seeing,' " *The Telegraph*, July 19, 2015, http://www.telegraph.co.uk/news/worldnews/africaandindianocean/tanzania/11748349/Tanzanias-elephant-catastrophe-We-recalculated-about-1000-times-because-we-didnt-believe-what-we-were-seeing.html.

238 Dr. Paul O'Donoghue, a British scientist working with rhinos: MacDonald, Fiona, "Scientists are Fitting Rhinos' Horns with Spy Cameras and Alarms to Catch Poachers," *Science Alert*, July 20, 2015, http://www.sciencealert.com/scientists-are-fitting-rhinos-horns-with-spy-cameras-and-alarms-to-catch-poachers.

239 "Ivory operates as [a] savings account for Kony: Christy, Bryan, "How Killing Elephants Finances Terror in Africa," *National Geographic*, August 12, 2015, http://www.nationalgeographic.com/tracking-ivory/article.html.

239 a carved tusk in China can go for $1,300: Scully, Matthew, "Inside the Global Industry that's Slaughtering Africa's Elephants," *The Atlantic*, June 6, 2013, http://www.theatlantic.com/international/archive/2013/06/inside-the-global-industry-thats-slaughtering-africas-elephants/276582/.

239 the US ambassador to Kenya noted "a marked increase in poaching wherever Chinese labor camps [are] located": Scully, Matthew, "Inside the Global Industry that's Slaughtering Africa's Elephants," *The Atlantic*, June 6, 2013, http://www.theatlantic.com/international/archive/2013/06/inside-the-global-industry-thats-slaughtering-africas-elephants/276582/.

239 "Poaching has risen sharply in areas where the Chinese are building roads . . .": Scully, Matthew, "Inside the Global Industry that's Slaughtering Africa's Elephants," *The Atlantic*, June 6, 2013, http://www.theatlantic.com/international/archive/2013/06/inside-the-global-industry-thats-slaughtering-africas-elephants/276582/.

240 There were 130,000 elephants in Sudan twenty-five years ago: Holland, Hereward, "South Sudan's Elephants Could be Wiped Out in 5 Years," Reuters Green Business, December 4, 2012, http://www.reuters.com/article/2012/12/04/us-sudan-south-elephants-idUSBRE8B30LN20121204.

240 "Sudanese hunting expeditions are today operating more than 600 km . . .": Vira, Varun, and Thomas Ewing, "Sudan: Failed States & Ungoverned Corridors," *Ivory's Curse: The Militarization & Professionalization of Poaching in Africa*, Born Free, April 2014, 24.

240 there were perhaps a million elephants in the Central African Republic alone: "Elephant Watch," *The New Yorker*, May 11, 2015. http://www.newyorker.com/magazine/2015/05/11/elephant-watch.

240 Today, there are probably fewer than 100,000 elephants in the entire group of countries in central Africa: "Elephant Watch," *The New Yorker*, May 11, 2015.

240 nearly a thousand park rangers have been killed worldwide: Smith, David, "Fighting the Poachers on Africa's Thin Green Line," *The Guardian*, June 15,

2013, http://www.theguardian.com/theobserver/2013/jun/15/poachers-africa-thin-green-line.

240 after Sudanese poachers ambushed and killed five rangers: Vira, Varun, and Thomas Ewing, "Sudan: Failed States & Ungoverned Corridors," *Ivory's Curse: The Militarization & Professionalization of Poaching in Africa,* Born Free, April 2014, 24.

240 In March 2015, Ethiopia's Deputy Prime Minister . . . set ablaze his country's entire stockpile of six tons of ivory: "Ethiopia Destroys Six-Tonne Stockpile of Poached Ivory," *BBC News,* March 20, 2015, http://www.bbc.com/news/world-africa-31983727.

240 President Arthur Peter Mutharika pledged to burn: "Malawi Gets 'Serious' About Wildlife Crime," Environmental News Service, August 9, 2015. http://ens-newswire.com/2015/08/09/malawi-gets-serious-about-wildlife-crime/.

240 "The destruction of the ivory stockpile . . .": Stansfield, Kat, "Malawi Takes to the Screen to Say No to Wildlife Crime," *Africa Geographic,* April 8, 2015, http://africageographic.com/blog/malawi-takes-to-the-screen-to-say-no-to-wildlife-crime/.

241 President Denis Sassou Nguesso presided over the burning of five tons of ivory there: Smith, David, "Congo-Brazzaville President Burns Five Tonnes of Ivory in Fight Against Poachers," *The Guardian,* April 29, 2015, http://www.theguardian.com/world/2015/apr/29/congo-brazzaville-president-burns-elephant-ivory-poachers.

242 The United States government, in an effort to discourage the purchase and sale of ivory: US Fish & Wildlife Service, "The U.S. Ivory Crush at Times Square," July 20, 2015, http://www.fws.gov/le/elephant-ivory-crush.html.

242 New York and New Jersey banned the trade in ivory: Baker, Peter, and Jada Smith, "Obama Administration Targets Trade in African Elephant Ivory," *The New York Times,* July 25, 2015, http://www.nytimes.com/2015/07/26/world/africa/obama-administration-targets-trade-in-african-elephant-ivory.html?_r=0.

242 President Obama announced a new policy: Gitau, Beatrice, "Obama Announces More Restrictions on Sale of Ivory from African Elephants," *The Christian Science Monitor,* July 25, 2015, http://www.csmonitor.com/USA/USA-Update/2015/0725/Obama-announces-more-restrictions-on-sale-of-ivory-from-African-elephants.

242 the Chinese government destroyed 662 kilograms of confiscated ivory: African Wildlife Foundation, "China Destroys More Ivory in Symbolic Gesture," May 29, 2015, https://www.awf.org/news/china-destroys-more-ivory-symbolic-gesture.

242 "We will strictly control ivory processing . . .": African Wildlife Foundation, "China Destroys More Ivory in Symbolic Gesture," May 29, 2015, https://www.awf.org/news/china-destroys-more-ivory-symbolic-gesture.

242 "enact nearly complete bans on ivory import and export . . .": Fact Sheet, "President Xi Jinping's State Visit to the United States," September 25, 2015, https://www.whitehouse.gov/the-press-office/2015/09/25/fact-sheet-president-xi-jinpings-state-visit-united-states.

246 "The blood flowed and it wasn't a pretty sight . . .": BBC News, "Bloody End

to Norway Whale Safari," July 6, 2006, http://news.bbc.co.uk/2/hi/europe/5153602.stm.

246 "should be a thing of the past . . .": Charles, Jacqueline, "Easter Caribbean Whalers Follow a 139-Year-Old Tradition, Now Under Siege," *Miami Herald*, April 16, 2014, http://www.miamiherald.com/news/nation-world/world/americas/article1962769.html.

247 Mountain gorillas are the most endangered great apes in the world: WWF, "Mountain Gorilla," n.d., http://www.worldwildlife.org/species/mountain-gorilla.

247 "There are no devils left in Hell . . .": Gibbs, Nancy, "Why? The Killing Fields of Rwanda," *Time*, May 16, 1994, http://content.time.com/time/magazine/article/0,9171,980750,00.html.

248 "to save the last great species and places on earth for humanity": Kenyan Wildlife Service, *About Us*, n.d., http://www.kws.go.ke/about-us/about-us.

249 the navy burned and sank six of them: Erdmann, Mark, "Indonesian Government Sinks Vietnamese Shark Poaching Boat, Creates New Dive Site," Humanature, February 10, 2015, http://blog.conservation.org/2015/02/indonesian-government-sinks-vietnamese-shark-poaching-boat-creates-new-dive-site/.

249 the navy sunk a large Vietnamese ship: Erdmann, Mark, "Indonesian Government Sinks Vietnamese Shark Poaching Boat, Creates New Dive Site," Humanature, February 10, 2015, http://blog.conservation.org/2015/02/indonesian-government-sinks-vietnamese-shark-poaching-boat-creates-new-dive-site/.

249 the annual value to the tourism industry of an individual reef shark: Vianna, G.M.S, M.G. Meekan, D. Pannell, S. Marsh, and J.J. Meeuwig, Executive Summary, "Wanted Dead or Alive? The Relative Value of Reef Sharks as a Fishery and an Ecotourism Asset in Palau," 2010, http://www.pewtrusts.org/~/media/Assets/2011/05/02/Palau_Shark_Tourism.pdf.

249 One of the biggest gathering places for the world's largest fish: Douthwaite, Karen, "Know Before You Go: Responsible Whale Shark Tourism in Mexico," August 30, 2013, http://www.worldwildlife.org/blogs/good-nature-travel/posts/know-before-you-go-responsible-whale-shark-tourism-in-mexico.

249 "The Yucatan's whale shark tourism industry has grown tremendously": Douthwaite, Karen, "Know Before You Go: Responsible Whale Shark Tourism in Mexico," August 30, 2013, http://www.worldwildlife.org/blogs/good-nature-travel/posts/know-before-you-go-responsible-whale-shark-tourism-in-mexico.

249 the Indonesian Council of Ulama issued a fatwa . . .: Christy, Bryan, "First Ever Fatwa Issued Against Wildlife Trafficking," *National Geographic*, March 5, 2014, http://news.nationalgeographic.com/news/2014/03/140304-fatwa-indonesia-wildlife-trafficking-koran-world/.

250 "This fatwa is issued to give an explanation . . .": Christy, Bryan, "First Ever Fatwa Issued Against Wildlife Trafficking," *National Geographic*, March 5, 2014, http://news.nationalgeographic.com/news/2014/03/140304-fatwa-indonesia-wildlife-trafficking-koran-world/.

250 an estimated two thousand baby orangutans were exported: Orme, David, *Animals Under Threat: Orangutan* (Chicago: Heinemann Library, 2005), 28.

CONCLUSION: **High Yield Bonds**

253 it passed . . . with more than 70 percent of the vote: "Initiative Measure No. 1401 concerns trafficking of animal species threatened with extinction," November 3, 2015 General Election Results, http://results.vote.wa.gov/results/current/State-Measures-Initiative-Measure-No-1401-concerns-trafficking-of-animal-species-threatened-with-extinction.html.

254 "They are aggressive . . .": Shoichet, C., "Freed Colombian Hostages Carry Pets Tamed in the Jungle," CNN, April 4, 2012, http://www.cnn.com/2012/04/03/world/americas/colombia-farc-hostages/.

255 "His name is Rango . . .": Shoichet, C., "Freed Colombian Hostages Carry Pets Tames in the Jungle," CNN, April 4, 2012, http://www.cnn.com/2012/04/03/world/americas/colombia-farc-hostages/.

257 Timothy Pachirat . . . spent six months working on the floor of a cattle slaughter plant: Pachirat, T., *Every Twelve Seconds: Industrialized Slaughter and the Politics of Sight*, Yale University Press, 2013. pp. 1–84.

257 "the police waved the workers back and opened fire . . .": Pachirat, T., *Every Twelve Seconds: Industrialized Slaughter and the Politics of Sight*, Yale University Press, 2013.p. 2.

258 Pachirat diagrammed the floor plan and identified the duties of the 120 people: Pachirat, T., *Every Twelve Seconds: Industrialized Slaughter and the Politics of Sight*, Yale University Press, 2013. pp. 38–84.

258 "They don't want to be reminded that it was a live animal": Matlack, C., "Why Did KFC Let a TV Crew Inside Its Chicken Operation?", *Bloomberg Business*, March 18, 2015, http://www.bloomberg.com/news/articles/2015–03–18/why-did-kfc-let-a-tv-crew-inside-its-chicken-operation-.

260 "all human sympathy to keep its distance": Dickens, C., *A Christmas Carol*, 1843.

261 "our treatment of animals . . . abominable by succeeding generations . . .": Krauthammer, C., "Free Willy!", *The Washington Post*, May 7, 2015, https://www.washingtonpost.com/opinions/free-willy/2015/05/07/4d1a82f2-f4f2–11e4-b2f3-af5479e6bbdd_story.html.

261 "Is making a hen's cage a little larger really so cost-prohibitive . . .": Parker, K., "Steve King's Inhumane Farm Bill Measure," *The Washington Post*, August 20, 2013, https://www.washingtonpost.com/opinions/kathleen-parker-steve-kings-inhumane-farm-bill-measure/2013/08/20/bf51f688–09d1–11e3–8974-f97ab3b3c677_story.html.

261 "We as a global society have crossed the Rubicon . . .": Kristof, N., "Can We See Our Hypocrisy to Animals?", *The New York Times*, July 27, 2013, http://www.nytimes.com/2013/07/28/opinion/sunday/can-we-see-our-hypocrisy-to-animals.html?_r=0.

262 "systematic animalizing of African Americans": Davis, D.B., *The Problem of Slavery in the Age of Emancipation*, 2014.

262 "Slaveholders regard their slaves not as human beings, but as mere working animals, as merchandise": Weld, T.D., *American Slavery As It Is*, Knopf Doubleday, 2015. p. 7.

262 "I can disappear behind Choupette": Scheiner, M., "Karl Lagerfeld on Fur (Yea), Selfies (Nay) and Keeping Busy," *The New York Times*, March 3, 2015, http://www.nytimes.com/2015/03/05/fashion/karl-lagerfeld-on-fur-yea-selfies-nay-and-keeping-busy.html.

262 "hate[s] the idea of killing animals in a horrible way . . .": Scheiner, M., "Karl Lagerfeld on Fur (Yea), Selfies (Nay) and Keeping Busy," *The New York Times*, March 3, 2015, http://www.nytimes.com/2015/03/05/fashion/karl-lagerfeld-on-fur-yea-selfies-nay-and-keeping-busy.html.

263 "we will not be using any farmed fur . . ." Ledbetter, C., "Hugo Boss Says It Plans To Go Fur Free By 2016," *HuffPost Style*, July 8, 2015, http://www.huffingtonpost.com/entry/hugo-boss-fur-free_559c31b5e4b0759e2b511c4e.

264 "Our indifference or cruelty toward fellow creatures of this world . . .": Pope Francis, "V. A Universal Communion," *Praise be to You—Laudato Si: On Care for Our Common Home,* 2015, Ignatius Press.

264 "We read in the Gospel that Jesus says of the birds of the air . . .": Pope Francis, "III. Ecological Conversion," *Praise be to You—Laudato Si: On Care for Our Common Home,* 2015, Ignatius Press.

267 According to a report from the University of Florida: Batz, Michael B., Hoffmann, Sandra, and Morris, Jr., J. Glenn, "Ranking the Risks: The 10 Pathogen-Food Combinations With The Greatest Burden on Public Health," Emerging Pathogens Institute of the University of Florida, 2011.

268 in the United States, "at least two million people become infected with bacteria that are resistant to antibiotics . . .": CDC, "Antibiotic/Antimicrobial Resistance," n.d., http://www.cdc.gov/drugresistance/.

268 "much of the use of antibiotics in animals . . .": CDC, "Fighting Back Against Antibiotic Resistance," *Antibiotic Resistance Threats in the United States, 2013*, 31, http://www.cdc.gov/drugresistance/pdf/ar-threats-2013–508.pdf.

268 "a problem so serious it threatens the achievements of modern medicine . . .": World Health Organization, Foreword, *Antimicrobial Resistance: Global Report on Surveillance, 2014*, IX, http://www.who.int/drugresistance/documents/surveillancereport/en/.

268 Industrial-style pig farmers around the world use, on average, nearly four times as many antibiotics: Van Boeckel, Thomas; Brower, Charles; Gilbert, Marius; et al: Proceedings of the National Academy of Sciences of the United States of America: "Global Trends in Antimicrobial Use in Food Animals," February 18, 2015 http://www.pnas.org/content/112/18/5649.abstract, and Doucleff, Michaeleen: National Public Radio, "For The Love Of Pork: Antibiotic Use On Farms Skyrockets Worldwide," March 20, 2015 http://www.npr.org/sections/goatsandsoda/2015/03/20/394064680/for-the-love-of-pork-antibiotic-use-on-farms-skyrockets-worldwide.

268 A number of major food retailers, notably McDonald's, Costco, and Walmart: NBC News, "Costco Wants to End Use of Human Antibiotics in Chicken," March 6, 2015, http://www.nbcnews.com/business/consumer/costco-working-end-use-human-antibiotics-chicken-n318366 and Strom, Stephanie; "Tyson to End Use of Human Antibiotics in Its Chickens by 2017," *The New York Times*,

April 29, 2015. http://www.nytimes.com/2015/04/29/business/tyson-to-end-use-of-human-antibiotics-in-its-chickens-by-2017.html?emc=edit_tnt_20150428&nlid=65739825&tntemail0=y&_r=0

269 At the US Meat Animal Research Center (USMARC): Aubrey, Allison, "Outrage Over Government's Animal Experiments Leads To USDA Review," *National Public Radio*, February 6, 2015, http://www.npr.org/sections/thesalt/2015/02/06/384103870/outrage-over-governments-animal-experiments-leads-to-usda-review.

270 Burmese pythons have wiped out 99 percent of raccoons, opossums, and other small and medium-sized mammals: Dorcas, Michael, et al., *Severe Mammal Declines Coincide with Proliferation of Invasive Burmese Pythons in Everglades National Park*, 2011, http://www.pnas.org/content/109/7/2418.full.

271 The estimated annual cost to the United States of removing and managing invasive species . . . is $120 billion: US Fish & Wildlife Service, *The Cost of Invasive Species*, 2012, https://www.fws.gov/verobeach/PythonPDF/CostofInvasivesFactSheet.pdf.

271 Kill buyer Dorian Ayache got into multiple accidents: Habitat for Horses, "Slaughter-Horse Hauler Enters Plea," 2014, http://www.habitatforhorses.org/slaughter-horse-hauler-enters-plea-video/.

272 In 75 percent of cases where there is animal cruelty at a home: American Humane Association, Facts About Animal Abuse & Domestic Violence, n.d., http://www.americanhumane.org/interaction/support-the-bond/fact-sheets/animal-abuse-domestic-violence.html.

273 The handle . . . declined from $3.5 billion in 2001 to $600 million today: Cook, Amelia, "High Stakes: Greyhound Racing in the United States," Grey2K USA and the ASPCA, 2015, p. 16.

273 twenty-two horses die on racetracks each week: Bogdanich, Walt, et al., "Mangled Horses, Maimed Jockeys," *The New York Times,* March 24, 2012, http://www.nytimes.com/2012/03/25/us/death-and-disarray-at-americas-racetracks.html?pagewanted=all&_r=0.

274 Perdue Farms affixed "humanely raised" to labels for its chicken: Perdue Will Remove Its "Humanely Raised" Label to Settle HSUS Lawsuits, *Food Safety News*, October 14, 2014. http://www.foodsafetynews.com/2014/10/perdue-to-remove-its-humanely-raised-label/#.VoiO5k3rvDc.

274 Of Americans with pets, 91 percent consider their pets to be family members: "More than Ever, Pets are Members of the Family," *The Harris Poll*, July 16, 2015, http://www.theharrispoll.com/health-and-life/Pets-are-Members-of-the-Family.html.

Index